New Prospects of Integrating Low Substrate Temperatures with Scaling-Sustained Device Architectural Innovation

Synthesis Lectures on Emerging Engineering Technologies

Editor
Kris Iniewski, *Redlen Technologies, Inc.*

New Prospects of Integrating Low Substrate Temperatures with Scaling-Sustained Device Architectural Innovation
Nabil Shovon Ashraf, Shawon Alam, and Mohaiminul Alam
2016

Compound Semiconductor Material and Devices
Zhaoiun Liu, Tondgde Huang, Qiang Li, Xing Lu, and Xinbo Zou
2016

Advances in Reflectometric Sensing for Industrial Applications
Andrea Cataldo, Egidio De Benedetto, and Giuseppe Cannazza
2016

Sustaining Moore's Law: Uncertainty Leading to a Certainty of IoT Revolution
Apek Mulay
2015

New Prospects of Integrating Low Substrate Temperatures with Scaling-Sustained Device Architectural Innovation

Nabil Shovon Ashraf, Shawon Alam, and Mohaiminul Alam

ISBN: 978-3-031-00899-3   paperback
ISBN: 978-3-031-02027-8   ebook

DOI 10.1007/978-3-031-02027-8

A Publication in the Springer series
SYNTHESIS LECTURES ON EMERGING ENGINEERING TECHNOLOGIES

Lecture #4
Series Editor: Kris Iniewski, *Redlen Technologies, Inc.*
Series ISSN
Print 2381-1412   Electronic 2381-1439

New Prospects of Integrating Low Substrate Temperatures with Scaling-Sustained Device Architectural Innovation

Nabil Shovon Ashraf, Shawon Alam, and Mohaiminul Alam
North South University

SYNTHESIS LECTURES ON EMERGING ENGINEERING TECHNOLOGIES #4

ABSTRACT

In order to sustain Moore's Law–based device scaling, principal attention has focused on toward device architectural innovations for improved device performance as per ITRS projections for technology nodes up to 10 nm. Efficient integration of lower substrate temperatures (<300K) to these innovatively configured device structures can enable the industry professionals to keep up with Moore's Law-based scaling curve conforming with ITRS projection of device performance outcome values. In this prospective review E-book, the authors have systematically reviewed the research results based on scaled device architectures, identified key bottlenecks to sustained scaling-based performance, and through original device simulation outcomes of conventional long channel MOSFET extracted the variation profile of threshold voltage as a function of substrate temperature which will be instrumental in reducing subthreshold leakage current in the temperature range 100K–300K. An exploitation methodology to regulate the die temperature to enable the efficient performance of a high-density VLSI circuit is also documented in order to make the lower substrate temperature operation of VLSI circuits and systems on chip process compatible.

KEYWORDS

threshold voltage, substrate temperature, Fermi potential, intrinsic carrier concentration, bulk potential, depletion charge, metal-to-semiconductor work function difference, flat-band voltage, subthreshold leakage current, thin-film microcoolers

Contents

CHAPTER 1

Review of Research on Scaled Device Architectures and Importance of Lower Substrate Temperature Operation of n-MOSFETs

1.1 INTRODUCTION AND SCOPE OF THIS E-BOOK

With regard to the current research impetus that has been prevalent in the device engineering and modeling arena, most technological denouements are accompanied by room temperature or 300K device architectural-based new alternative and emerging device structures to the mainstream CMOS-based devices. The key goals in pursuit of these alternative device structures are to improve the off-state leakage current (I_{off}), superior I_{on} (on-state drive current)/I_{off} ratio, enhanced inversion layer mobility and steep subthreshold slope (less than 60 mV/decade even at room temperature 300K). Various novel device structures such as Ge-on—Silicon-on-Insulator (GESOI), staggered heterojunction vertical Tunnel FET, III-V MOSFET and Tunnel FET, Gate-all-around (GAA) nanowire tunnel FET, etc., are the competitive device architectures that are being vigorously employed to sustain Moore's Law-based scaling with simultaneous accordance with the key goals or benchmark figures of merit mentioned above. The process and fabrication complexities associated with these device architectures are proving to be very challenging for manufacturing professionals to keep the device or substrate temperature at 300K yet achieve the subthreshold slope lower than 60 mV/decade, the room temperature limit of thermal velocity related carrier surmounting of source side barrier of a natural MOSFET. Tunnel FET exhibits a room temperature subthreshold slope which is less than 60 mV/decade by employing different transport features to inject carriers into the channel and does not need thermal velocity related excitation to emit the carriers over the source side barrier. After a close examination and forethought into these exciting developments that have come about only in the last four or five years, the authors of this E-book felt the urge to pursue reduced (less than 300K) substrate or lattice temperature operation of present day MOSFET structures and other alternative device structure examples mentioned above to keep pace with Moore's Law-based scaling curve and achieve the key goals mentioned above in a more efficient way, with fewer device-centered and process-centered hindrances or bottlenecks encountered during batch manufacturing and on-chip systems integration and circuit performance. Even though the concept of low temperature electronics or cryo-cooling was developed much earlier and researchers from IBM have already analyzed device performance based on low temperature cooling, one reason the industry has been hesitant to incorporate the cooling integrated device structures with silicon dies is because of the cost associated with the cooling need which will negatively impact the consumer markets. With this context, the authors of this E-book opine that even the room temperature 300K alternative and emerging MOSFET structure such as the GAA III-V tunnel FET requires many process sequences in ultimate fabrication, which has already proved to be highly expensive compared to the required price per chip when ULSI systems-on-chip are considered. Instead of operating the device at 77K or at liquid helium or nitrogen temperature that most earlier research on low temperature electronics concentrated upon, the device substrate temperature can be tailored to operate at 250K or 200K which is not a substantial reduction from 300K, and the overhead for cooling related solid state ICs to be integrated with silicon die will not be substantial when the substrate temperature has to be within 100K of room temperature. With this implementation of reduced substrate temperature, the E-book systematically discusses the different modeling features of threshold voltage to eventually enable the device engineers to attain a host of performance benchmarks that they are investing in emerging device structural solutions, albeit at 300K operation. The authors thus feel that this E-book will be very useful to the readers or engineers concentrating on device physics-based analyses of lower substrate temperature operational benefits that are first extracted from long channel MOSFET device features and extended to today's short channel-based device architectures following a scaling ratio similar to constant field-based scaling theory employed for MOSFET scaling. At the end of the composition of this E-book, diverse set of richly blended list of references is enumerated for the devoted and inspired perusers to meditate and garner relevant facts and information which have been principally guiding impetus behind most of the contents of this E-book particularly the write-ups in the relevant sections on device architectural issues sustained scaled operation of MOSFETs and previous researches conducted on low temperature electronics, i.e., the breakthrough mainstream theme of this E-book.

1.2   BASIC OVERVIEW AND OPERATIONAL SALIENT FEATURES OF n-CHANNEL MOSFET DEVICE TRANSPORT

Although the CMOS device is the core element of every integrated circuit building blocks of microprocessor logic and memory units, most textbooks and research articles focus of n-channel MOSFET because electrons are the channel carrier with higher intrinsic and surface mobility and also since for n-MOSFET, if we consider the enhancement mode type operation, positive gate voltage bias is needed for channel conduction. It is easier to design power supply blocks for generating positive supply voltages rather than negative supply voltages required for channel conduction in *p*-channel MOSFETs where the transport is governed by holes, including the fact that holes have intrinsic mobility 2.5 order lower than electrons as carriers. Since this E-book details the scaling-based operational characteristics' improvements of *n*-channel MOSFETs, we would like to briefly document some of the salient device physical analyses on *n*-channel MOSFET turn-on and turn-off characteristics with effects of substrate doping, substrate bias, and oxide thickness and how the device depletion width profile evolves from long channel regime to short channel counterpart. As shown in Figure 1.1, the metal-oxide-semiconductor (MOS) field-effect transistor (FET) structure can be intuitively configured from the concept of the metal-oxide semiconductor (MOS) capacitor. In *n*-MOS capacitor structure, the substrate is *p*-type and grounded.

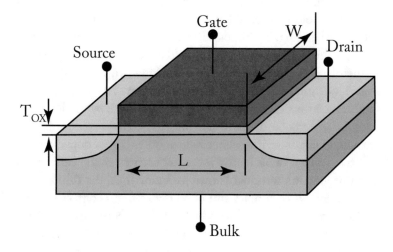

Figure 1.1: MOSFET structure.

A thin oxide is grown on *p*-type substrate and on top this oxide layer metal coating is overlaid to configure the gate terminal. The top metal terminal is called the gate terminal as the bias voltage applied to it, the thin inversion channel (*n*-type), can form underneath the oxide layer at $Si:SiO_2$ interface or the interface can be depleted of *n*-type carriers or accumulated by *p*-type hole carriers. When negative voltage is applied to top metal contact of this MOS capacitor from the point of flat band voltage, holes get accumulated to the surface attracted by the negative potential on the gate and the negative bias thus constitutes the accumulation region. But these mobile carriers are not minority carriers, such as electrons in *p*-type substrate that can be isolated from the bulk by the depletion region and utilized for conduction at the surface. Rather the accumulation regime signifies a redistribution of majority holes from the top surface to the device bulk. From the accumulation regime as the gate voltage is increased beyond flat band voltage toward more positive polarity, the positive charge on the gate repels the mobile holes at the surface and they are pushed deeper toward the bulk, leaving the immobilized negatively charged depletion ions near the Si-SiO_2 surface forming the depletion region as the region is depleted of mobile p-type holes. If we continue applying increasing positive voltage on the gate, the depletion width expands and minority carriers such as electron-hole pairs get generated in the depletion region, and the electron carriers thus generated by thermal excitation get attracted toward the surface and start forming thin inversion charge density. At a certain positive gate voltage for n-MOS capacitor device, the depletion width reaches its maximum and the inversion charge concentration at the Si:SiO_2 interface becomes equal to the majority acceptor dopant concentration through at the bulk, clearly establishing the condition that the surface has been "inverted" or the n-type minority carrier at the bulk has just become the n-type majority carrier at the surface. The gate voltage which establishes this condition is known as the threshold voltage. As the gate voltage is increased incrementally over threshold voltage, there is almost no addition to the depletion region by the additional surface band bending, but almost all of it goes into increasing the inversion charge density at the surface. At this point the concept of field effect transistor can be visualized. If for this gate voltage on the metal contact maintaining inversion charge density at the Si:SiO_2 interface we can add a source junction with significant concentration of n-type carriers, we can inject electrons and maintain this inversion charge density at the surface by modulating the metal-contact on the oxide layer, i.e., gate potential to source junction contact voltage difference. On the other side of the semiconductor surface if we can create another highly doped n-type junction and put enough high positive voltage upon it, the inversion carriers at the surface will move out or be "collected" out of this junction terminal known as drain. Thus, the thin inversion charge density is constantly supplied by source contact and flows out of drain contact—essentially forming the concept of MOSFET (metal-oxide-semiconductor-field-effect-transistor). The switching time of MOS capacitor and MOSFET is orders of magnitude difference since in the MOS capacitor the initial formation of inversion charge density takes time because of a thermally activated inversion charge or electrons in the depletion region, but in the case of MOSFET, these inversion carriers are injected and maintained by the source contact being ohmic and controlled by gate-to-source potential difference and hence take considerably less time in generation of minority carriers. Now we focus on how gate-to-source voltage controls the injection of minority electrons into the channel, i.e., the thin inversion layer adjoining source and drain n^+ contact regions. First of all, there is a reverse bias region between n^+ source and p-type body, and based on body doping and rather degenerate n^+ doping a built-in voltage is formed across the source-to-body reverse biased diode. This reverse biased diode ideally inhibits carrier injection into the channel from the source at zero voltage on the gate and small positive or negative voltage. Therefore, on the energy band diagram, the gate-to-source barrier is very high so that with the acquired thermal or drift velocity it will be almost impossible for carriers to surmount this barrier. As the gate voltage is increased with the source at ground potential, this voltage gets coupled to the reverse biased depletion region of the source-to-body junction close to the source and reduces the built-in voltage of the source-to-body reverse biased diode. As a result of this, the gate-to-source barrier is progressively reduced and at threshold, the barrier is of such height that the carriers easily surmount the barrier through their thermal velocity and populate the channel from the source side. Another concept that is misinterpreted is the reference potential by which gate voltage difference or drain voltage difference is computed. Ideally the p-type substrate contact potential is taken as reference and mid-channel potential is considered as the reference point for measuring threshold voltage or maximum depletion width. Proper knowledge out of device physical principles tells us since the source controls the injection of carriers into the channel, all references must be made with regard to source potential, including computations of threshold voltage, and maximum depletion width with regard to surface band bending potential must be referred from the source junction. Also at long channel regime, when there is ideal spacing between source and drain junctions, a thin oxide ensures proper termination of gate field lines, and the depletion region profile is rather flat for most of the gate bias ranges for small drain biases. Under this condition, since the depletion region profile is flat from source to the drain in the channel, maximum depletion width will be spatially non-variant from source to the drain, and mid-channel reference will not alter the correct computation of threshold voltage. But as we enter the short channel regime, 2D field effects both from gate and drain terminal make the depletion region profile highly non-uniform from source to drain, having a peak near the middle of the channel within a small range. At this condition, calculation of maximum depletion width in the mid-channel region will introduce error as the source side maximum depletion width extension will be critical here to correctly compute the threshold voltage. A four-terminal MOSFET is equipped with additional terminal at the body to modulate the carrier transport in the channel of MOSFET. A negative bias on the p-type substrate contact will increase the reverse bias width of the source-body reverse bias diode and transistor threshold voltage, for perfectly "on" characteristics will shift toward a higher value. When 2D effects or potential contours become important in short channel MOSFETs, application of this negative bias on p-type body is crucial in controlling the non-uniform depletion width profile from source side into the channel toward the drain side. A positive bias voltage on the body on the other hand forward biases the source-to-body junction diode and hence threshold voltage for conducting channel will be set at a lower gate voltage than optimum. A positive voltage on the bulk will also reduce the vertical field from the gate and the reverse bias depletion field from the drain, making the 2D effect less severe for short channel MOSFET operation, essentially smoothing the asperities on the depletion width edge profile from source to drain. Interested readers are hereby directed to some excellent text books available for institutional instruction and research purposes covering MOSFET device physics and carrier transport [1–14].

1.3 REVIEW OF CHALLENGES AND BOTTLENECKS EXPERIENCED OVER SUSTAINED MOSFET DEVICE SCALING

Inquisitive readers, who may comprise technology professionals, researchers, educators, and pupils from around the world, may be interested in some classic references cited in order comprising of original research articles particularly on device scaling based architectural aspects and their robust design schemes [19–38]. We begin the discussion in this section by citing the insightful excerpts from the classic textbook *Fundamentals of Modern VLSI Devices*, edited by Y. Tuar et al. [4]. A CMOS circuit typically consists of an n-channel MOSFET and a p-channel MOS-FET connected in series between the power supply terminals so that when the circuits are driven inactive (stand by condition), there is negligible standby power dissipation. Significant power is dissipated only during dynamic activity (simultaneous switching of n- and p-MOSFET device core and also for dynamic logic or pseudo-logic device core) or switching of the circuit. By ingeniously designing the "switching activities" of the circuits on a chip to minimize active power dissipation, engineers have been able to integrate hundreds of millions of CMOS transistors on a single chip and still have the chip readily air-coolable. Until recently, the integration level of CMOS was not limited by chip level power dissipation but by chip fabrication technology. Another advantage of CMOS circuits comes from the ratio-less full rail-to-rail logic swing, which improves the noise margin even practically constrained by the requirement of low voltage supply operation, and this unique advantage makes a modular cell based CMOS chip easier to design. As linear dimensions reached the $0.5\,\mu m$ level in the early 1990s, the performance advantage of bipolar transistors was outweighed by the significantly greater circuit density of CMOS devices fostered by concomitant reduction of manufacturing cost per chip. The system performance benefit of integrated functionality superseded that of raw transistor performance, and practically all the VLSI chips in production today are based on CMOS process technology. From the processing and manufacturing aspects, advances in lithography and etching technologies to enable resolution of critical dimension (CD) of gate length of CMOS have guided the industry to scale down transistors in physical dimensions and to pack more transistors in the same chip area. Such progress, combined with a steady growth in chip size, resulted in an exponential growth in the number of transistors and memory bits per chip. Dynamic random access memories (DRAMs) have characteristically contained the highest component count of any IC chips. This has been so because of the small size of the one transistor memory cell based memory architecture and because of the large and often insatiable demand for more memory in computing systems. One remarkable feature of silicon devices that fuels the rapid growth of the information technology industry is that their speed of propagation increases and their manufacturing cost decreases as their device feature size is reduced. This is one of the principal precursors of scaling induced device architectural evolution—a mainstream topic where much discussion has been laid out by noted researchers, industry collaborators, and institutional educators from the past to date and covered again in this introductory section of the E-book. The transistors manufactured today are 20 times faster and occupy less than 1% of the complete die area of those built 20 years ago. For instance, the increase in the clock frequency of microprocessors is the result of technology, logic, and device innovational schemes synergized into microprocessor core architectures through years of accumulative research built upon a single MOSFET device architecture engineering, enabling today's terahertz transmission of binary bits in its simplest form.

In lieu of attempting to critically analyze some of the classic and epochal research articles that have fueled the interests of device and manufacturing research professionals over the years, the authors of this E-book are urged by the need of well-rounded documenting of curving out novel and educational features from a list of references, authored by highly illustrious research professionals whose articles have served the purposes of CMOS device-based instruction in research institutions and also provided the pathway for the global semiconductor industry to benchmark CMOS digital and analog VLSI performance through device, process (front-end-of-line), and packaging (back-end-of-line) innovations. In this connection, the authors recommend the excellent review articles co-authored by H.S.-P Wong et al. on (i) Nanoscale CMOS [22], (ii) Device Scaling Limits of Si MOSFETs and their Application Dependencies [20], and (iii) Beyond the Conventional Transistor [23]. For the ease of understanding of the readers, the authors of this E-book are encouraged to chronicle the remaining subsections of this introductory section on the basis of a recent review article by Subhra Dhar et al. [33]. We also refer the reader to references [20, 22] and [23] to better supplement the discussion enumerated in [33]. The article by Subhra Dhar et al. on Advancement in Nanoscale CMOS Design En-Route to Ultra-Low Power Applications uses technologically sound and comparatively relaxed demonstration of central concepts behind scaling based device architectural evolution with relevant focus on performance parameters optimization to enable chip scaling. While discussing the review contents of this reference [33], the authors of this E-book would also investigate and justify the use and modeling of substrate and lattice temperature to corroborate such a scaling scenario, and where reduced substrate temperature operation of an entire chip will deservedly serve as phenomenal panacea to inconsistencies and discrepancies that might originate at a relevant technology node, due to performance bottlenecks [see 29, 30, 40–42, 54, 61–63]. The goal of operating the entire chip at reduced substrate and lattice temperature is to preserve the architectural innovation based device performance improvements but also improve intrinsic device reliability and life-time failure. When intrinsic parametric benefits aided by lowered temperature operation duly get integrated into the transport characteristics of devices as coming out of evolutionary architectural design, significantly higher performance metrics can be accrued and mastered that would not have been otherwise possible by keeping the doorway of architectural innovation open but excluding thermal engineering features from the chip. We now critically document the illustrations of article [33] in terms of device performance optimization, sustaining scaling and device attributes that need constant manipulation and control.

The process parameters in low power design are channel length, oxide thickness, threshold voltage, and doping concentration in the channel that fundamentally contribute to device performance. The low-power design technique should be such that it is cost sensitive to the process parameter variations. As technology scales down, the variations of these device-performance centric parameters are expected to be significantly disparate in future generations. The variations of leakage power and propagation delay in the transistors on a given die embedding a particular core architecture are different for different low power design methodologies applied to different multi-core ASIC and digital circuits. The role of threshold voltage (V_{th}) and subthreshold swing (S) has become increasingly demanding with VLSI applications emphasizing ultra-low voltage subthreshold logic combined with low power high speed design. The gate oxide thickness in recent process technologies has approached the limit where the direct tunneling carrier electrons to the gate terminal starts to play a significant role in both off-state and on-state MOSFET transistor operation modes. This phenomenon, in addition to subthreshold leakage, results in a dramatic total static leakage power dissipation; for instance, the reference [44] addresses this issue. Thus, better design strategies to control the total leakage power are necessary. It is well know that gate tunneling currents are highly sensitive to the voltage variation across the gate oxide. Supply voltage attenuation commensurate with scaled die operation can give significant reduction in gate leakage power consumption. What has been apparent to the device and circuit designers is that traditional and conceptual circuit design techniques to mitigate the impact of gate leakage, which when provisioned will add to the chip area, would also be much less efficient than the use of alternative high-k materials in the standalone device fabrication itself. The fluctuations of oxide thickness variations in atomic scale result in exponentially diversified gate leakage current values whose cumulative effects from various circuit lay-outs of the die will far more outweigh the relative advantages deduced from reduced supply voltage (V_{DD}) and multi-threshold voltage CMOS operations based on scheduling the sleep mode operation of the circuit. This is the principal reason behind the use of high-k or higher-K dielectrics alternatives to SiO_2 in the device fabrication itself that will cut the gate leakage power at a faster rate than the circuit designers can demonstrate by the various circuit leakage power reduction methodologies [35]. With every technology scaling, as the device length between source and drain junction is also scaled proportionately, substrate or body doping needs to be increased proportionately to control the source side and drain side depletion extensions or encroachments into the substrate underneath the channel. Higher substrate doping results in degraded subthreshold swing, impeding the acceptable supply voltage scaling essential for acceptable standby power dissipation. Furthermore, supply voltage scaling (V_{DD}) necessitates reduction of threshold voltage (V_{th}) which exponentially increases the off-state leakage current I_{off}. I_{off} reduction is critical where the chips are often in standby (sleep) mode of operation and even during fractional clock cycle when the circuit is active, an acceptable I_{on}/I_{off} ratio still needs to be maintained since leakage power consumption is rapidly increasing at a much faster rate compared to dynamic power. Some of the front-end-of-line (FEOL) challenges faced by device designers in the deep submicron and ultra deep submicron regime of MOSFET operations are (i) diminishing $V_{gs} - V_{th}$ overdrive to sustain voltage scaling, (ii) larger V_{th}/V_{dd} where V_{th} scaling is much slower compared to V_{dd} from one technology node to the next, (iii) ultra shallow junctions necessitating dopant levels reaching the maximum solid solubility limit, and (iv) dopant loss due to inefficient activation anneal of source and drain side junctions owing to an extremely stringent thermal budget where rapid thermal annealing is employed, (v) increase of threshold voltage variability amongst intra-die and inter-die across a wafer due to statistical dopant fluctuations on very small geometry device area MOSFETs, (vi) controlling drain-induced-barrier-lowering (DIBL) through increasing substrate doping adversely affecting the channel electron mobility through increased Coulomb and surface roughness scattering due to enhancement of vertical gate field, (vii) higher substrate doping exposing too much sensitivity to substrate biases that are routinely used to upshift or downshift the nominal threshold voltage (V_{th}) of a circuit, and (viii) the random fluctuations of gate oxide thickness, making oxide scaling to less than 1.5 nm, almost unfeasible compared to other parametric scaling and beyond 1 nm limit; self-assembly of monolayers and atomic layer deposition (ALD) technique nevertheless highly resourceful still make the targeted conformal thickness of gate oxide deviate from optimal precision.

1.4 DEVICE PARAMETERS CRITICAL FOR PERFORMANCE ENHANCEMENT FOR GENERALIZED SCALING AND AT THE END OF MOORE'S LAW

We again critically document the reference paper [33] in relation to this subsection. Device scaling is based upon simple principles: by reducing the sizes of devices and interconnects, the packing density, power density, clock speed, and on-current performance of transistors can be improved. The principal effect with technology scaling is centered upon MOSFET channel length, where the generalized scaling factor is deduced from the reduction trend of channel length as adopted by ITRS and global semiconductor researchers, professionals, and manufacturers who also support Moore's Law-based scaling as company giant Intel Inc. emerges to be the largest proponent of Moore's Law-based scaling in its every technology node production of consumer products. This scaling scenario, stemmed originally from Dennard et al. papers on scaling theory [25–28], is based on the principle that when the dimensions of a MOSFET are scaled down, both the voltage level and the gate oxide thickness are also scaled. The supply voltage V_{dd} has been scaled down in direct proportion to channel length reduction in order to keep standby leakage current under control which has a direct bearing on power consumption. The transistor threshold voltage V_{th}, although does not scale in direct proportion to channel length reduction, does follow a reduction of 70.7% of scale factor and must be reduced to maintain a high drive current by making the gate overdrive factor nearly constant through scaling nodes, and for high enough drift and mobility of channel carriers, V_{th} reduction also maintains high enough inversion charge in the channel even when gate voltage is reduced. The presence of high enough inversion charge density enhances Coulomb field related mobility at a lower vertical gate field through dopant screening. In a given technology generation, since the source-body and drain-body depletion widths are predefined based on the doping, the rate at which barrier increases as a function of distance from the source into the channel is constant. As the channel length is further reduced, the uniformity of mid-channel potential is disturbed by the higher penetration fields of the drain potential entering laterally into the channel and becoming denser toward the surface channel region between source and drain. This drain field when terminating on neutral source junction reduces the gate controlled barrier at the junction, essentially injecting more electron carriers from source contact edge into the channel. This feature, in addition to short channel MOSFET scaling scenario, reduces the threshold voltage at a much steeper rate than conventional scaling theory will predict. Therefore, extra precaution and maximum variation of threshold voltage window must be stipulated by the device designers to ameliorate the excessive subthreshold leakage that results from steeper threshold voltage lowering, violating the design of minimum allowable threshold voltage roll-off from a circuit operational perspective.

The gate material has long been polysilicon with SiO_2 as the insulator between the gate and the channel. Aggressive scaling of CMOS technology in recent years has reduced the silicon dioxide (SiO_2) gate dielectric thickness below than 2 nm. Gate oxide thickness has direct bearing on gate integrity of channel, and thinner gate oxide empowers the gate potential to primarily control the channel charge distribution and channel potential even when low to moderate drain bias is present. As a result, reduction of gate oxide enables the gate to turn on the inversion channel faster, maintain high I_{on}/I_{off} ratio, and also turn off the inversion channel faster—a necessary ingredient for higher transistor on—current and higher switching speed. The device engineers found it much more practical to sustain the scaling scenario down to 25 nm gate length with reduction of gate dielectric in line with scaling factor, and transition to higher-k materials has only become practical in the last five years due to excessive gate tunneling leakage currents at the lower gate oxide thickness in the vicinity of 1.5 nm [59, 64–69]. It is also necessary to enhance the body doping to an almost degenerate condition in order to ensure MOSFET operation at a substantially reduced channel length to prevent a source-to-drain direct punchthrough current. At such excessive substrate doping, a subthreshold slope degrades as the depletion region capacitance is enhanced due to much thinner depletion width. Also band-to-band tunneling (BTBT) current becomes significant at source to body and drain to body junctions when substrate doping becomes of the order of source and drain junction doping values. BTBT results as the band bends so sharply at the depletion regions of source-to-body and drain-to-body junctions that electrons from the valence band can tunnel into empty conduction band via the narrow width created by a steep band bending well, and this can happen when the gate potential is low or intentionally kept negative for an n-MOSFET device to facilitate efficient off-current bias conditions. BTBT enhances gate induced drain leakage (GIDL) primarily at negative to zero gate voltage, and low to moderate drain voltage as drain voltage enhances the GIDL current by electron tunneling, by increasing reverse bias induced more band bending at the drain-to-body junction. Finally, excessive substrate doping reduces the amount of inversion charge density at a nominal reduced gate voltage and therefore Coulomb scattering can be significant. Even when the gate induced vertical field through the oxide is low, the excessive body doping can impart substantially surface potential bending which promotes interface roughness scattering through the oxide field, and this reduces mobility at a much faster rate than Coulomb scattering-related mobility reduction.

When V_{dd} is reduced toward shorter channel lengths, it becomes increasingly difficult to satisfy both the performance and off-current requirements, i.e., maintaining the required I_{on}/I_{off} ratio. Trade off between leakage current and circuit speed emanates due to subthreshold slope factor nonscalability which is intrinsically concatenated to device or substrate temperature reduction. Therefore, only the gate oxide thickness was reduced in due proportion to scaling factor but as we have entered the nanometer regime in the last 7–8 years, supply voltage V_{dd} has not been scaled down in proportion to channel length L and V_{th} has not been scaled in proportion to V_{dd}. This resulted in nearly constant or increased V_{th}/V_{dd} ratio degrading propagation delay and exacerbating switching speed. Also due to scaling, the contact area of source and drain junction along with shallow junction depth all are reduced and as a result the source, drain contact resistance, source to channel overlapping resistance, the shallow junction spreading resistance all are enhanced and impede speed of the device and the drive current level by imparting parasitic effects through RC delay where gate capacitance, gate side-wall capacitance, gate to junction spacer capacitance all get coupled to parasitic source and drain resistance contributing agents and degrade the device performance and AC gain whose impacts become mostly visible at high frequency, switching analog and mixed signal applications. For more insightful and resourceful feedback on the content of this subsection, see [43–58, 60].

1.5 ROLE OF SUBSTRATE TEMPERATURE MODELING AND CONTROL

This particular section sets the platform for the motivation and impetus behind the contents of this E-book and the readers are directed to the reference [24] of Yuan Taur on CMOS Design near the Limit of Scaling—particularly his expert narration of a projected outlook on extending CMOS scaling to 10 nm. From the scaling scenario that has enabled the improved device architecture at the 10 nm era, the principal drawback results from decreased I_{on}/I_{off} ratio, decreased effective mobility, and subthreshold slope nonscalability. Through the introduction of multi-gate device architecture and the Tunnel-FET structure, a subthreshold slope closer to a 60 mV/decade limit at $T = 300$K and even lesser of the order of 30–40 mV/decade for Tunnel FET structures have been achieved. But these devices, particularly TFET, still suffer from much lower I_{on} value of their conventional MOSFET counterpart and hence a much reduced I_{on}/I_{off} ratio where I_{off} is decades of magnitude lower than conventional silicon MOSFET, but at the same time I_{on} is also 3–4 orders of magnitude smaller than conventional silicon MOSFET. Even with a combination of strain and stress enhancing integration of materials in the channel and capping material on gate, the mobility slightly improves and still tends to be dominated by interface roughness scattering due to a vertical field that exists across the 1 nm scaled gate oxide even with thicker high-K gate stacks; as for successful integration of high-k materials with silicon substrate, interfacial sub 1 nm gate oxide cannot be fully removed or scavenged. Oxide tunneling nearly becomes unavoidable even with multigate nanowire architecture; if not we try to implement the intrinsic attributes of the device and integrate the benefits that we accrue from intrinsic parameter engineering such as substrate temperature with the device architectural based parametric innovations. The impetus for this book hinged on ingenious optimization and reduction of substrate or lattice temperature often associated with chip cooling that will enable sustained device performance even when the ultimate deciding factor, End of Moore's Law seems attractive. As the article [24] clearly states, the benefits are derived primarily from two aspects of MOSFET characteristics—higher carrier mobility and steeper subthreshold slope. According to the paper by Shin-ichi Takagi et al. [70], electron mobility enhances by a factor of more than 3 times from 300K to 77K substrate temperature depending on the magnitude of the vertical field and also when the body doping is less or near intrinsic. Similarly, the hole mobility improves by a factor of more than 2.5 times for the same temperature range. In addition, the MOSFET subthreshold slope steepens at a factor inversely proportional to the lattice temperature, making it much easier to turn off a device at lower than room temperature operation. Also at near intrinsic body doping where the impact of substrate temperature is maximum, the depletion region silicon capacitance (C_D) decreases due to wide enough depletion width, and depletion capacitance (C_D)/oxide capacitance (C_{ox}) decreases and this gets coupled to standard thermal energy based reduction (contribution from kT term), making subthreshold slope more steep through this additional beneficial contribution apart from simple proportional reduction of T (temperature). At lower than room temperature operation, the intrinsic carrier concentration of Si MOSFET reduces by orders of magnitude, and as a result the bulk potential and surface band bending at low enough gate voltage can still be substantially large, making the threshold voltage at a lower temperature larger than room temperature value. This has tremendous implications on the short channel behavior of the MOSFET. First for room temperature operation of MOSFET optimized for below 25 nm regime, absolute or nominal value of threshold voltage is already scaled and lowered but scaling induced 2D effects from drain, quantum mechanical effects from subbands, fringe fields from the sidewalls, higher-K gate capacitance all reduce the V_{th} even further, essentially making the V_{th} variability window difficult to track across die to die in a fabricated wafer. On the contrary, if the threshold voltage (V_{th}) at a reduced substrate temperature is shifted to a higher value than room temperature V_{th}, even with accommodation of considerable V_{th} variability window ΔV_{th} including the factors responsible for short channel effects, the final V_{th} that is measured across die-to-die will still be low enough as permitted by scaling but not too low to fail the V_{th} (min) window edge of V_{th} variability. One of the principal drawbacks at low temperature operation of MOSFET is the slower or retarding scaling of maximum depletion width W_{dm} with channel length L. As the channel length is scaled, for room temperature MOSFET, substrate doping has already been determined to be high enough to sustain scaling, and intrinsic carrier concentration is 10^{10}/cm^3. This makes the bulk potential considerably smaller compared to the gate voltage, and as the depletion width is directly proportional to the square root of the bulk potential effect at the surface band bending and inversely proportional to the square root of substrate doping, it is the substrate doping compared to bulk potential which reduces depletion width further as technology scales. For substantially low temperature operation as may be needed to enhance V_{th} to reduce off state leakage current and produce steeper subthreshold slope, this exposes the maximum depletion width W_{dm} nonscalability with channel length at a T around 100K. This is due to the reason that surface potential becomes non-varying and large at a lower T, sometimes larger than applied low gate voltage, owing to saturating $\ln\left(\frac{N_A}{n_i}\right)$ effect where N_A is of the order of 10^{16}–10^{18}/cm^3 as device scales from

1 micron to a few 10 nm but n_i (intrinsic carrier concentration of silicon) can be so low that careful calculation at a T around 100K, n_i value is less than 1. Hence, $\ln\left(\frac{N_A}{n_i}\right)$ reaches its maximum saturating limit incompatible with supply voltage reduction. The inversely proportional contribution to W_{dm}, square root of substrate doping alone cannot reduce W_{dm} in the adequate proportion such that the short channel efficiency factor $L_{eff} > 2W_{dm}$ is satisfied. Additionally, one device physical attribute at low temperature operation of extrinsic semiconductor is that incomplete ionization of acceptors of n-channel MOSFET becomes more prominent at lower T, and as depletion width scales inversely with doping value, it gets further boosted by the effect of incomplete ionization at every technology node when all the device designers will require is substantial reduction of maximum depletion width W_{dm} to enable gate length scaling. Therefore, this nonscalability feature of maximum depletion width at a sufficiently low substrate temperature of operation must be taken into account. By selecting a substrate temperature sufficiently close to room temperature where the advantages of improved carrier mobility and subthreshold slope factor can still be realized, use of low substrate doping can still reduce $\ln\left(\frac{N_A}{n_i}\right)$ considerably so that a moderate and in-line scaling of W_{dm} can still be possible. An intrinsic or near intrinsically doped n-channel MOSFET is being designed today to control V_{th} adjustment by metal gate work function engineering with the additional benefit of higher effective channel mobility. A near intrinsic channel in conjunction with not too low substrate temperature below $T = 300K$ operation of MOSFET can keep the W_{dm} scaling possible to avert the short channel effects. At the lower end of substrate temperature reduction, depletion width below the gate enhances at a faster rate which is obviously a concern for scaling enabled device architectural innovation. But since the depletion region near the drain also widens to include the increased surface band bending at lower T due to additional drain bias, drain can ideally lower the turn on V_{th} and cancel out the effect of increased W_{dm} by providing on current performance even at the expense of increased depletion width considering scaled operation and reduced substrate temperature. Non-uniform lateral doping where the doping is abruptly high near the source region and gradually decreases toward the channel up to the drain can ideally potentially reduce the overall maximum depletion width under the gate, and this reduction will be more effective as the source to drain distance narrows or constrict to a few atomic distance sustaining scaled operations. If the source to drain distance is of the order of a few nanometers, the variational effect on depletion layer width due to lateral non-uniform channel doping will be minimal since at the scaled operation of MOSFET, source side depletion width thickness is mostly relevant compared to mid-channel thickness, which is generally used in modeling equations for extracting various parameters owing to long channel and sub quarter micron channel length technology. Also inverse retrograde doping in the vertical direction (High-low) from channel surface to a few nm distance into the body can also alleviate the problem associated with depletion width enhancement at lowered substrate temperature operation. Inverse retrograde doping is engineered vertical non-uniform channel doping where the doping is steeply high at the channel surface to a few nm channel thickness that mostly embody the pathway for carrier flow from source to drain and then the doping is gradually reduced to average or effective body doping

along the vertical direction toward the bulk where surface potential bending gradually ceases to exist. Since the surface doping is high, the source side depletion region in two dimensions will include the effect of both steeply high vertical substrate doping and steeply high lateral substrate doping. Therefore, the extent to which a depletion region widens will be significantly curbed to represent the new modified W_{dm} that will still be sufficient to control the short channel and 2D effects. Substrate doping engineering in this way will therefore neutralize the impact of W_{dm} for scaled technology node operations of MOSFET at the lower substrate temperature condition.

Now that we have enumerated the efficient operation of MOSFET at the scaling limit, taking advantage of reduced substrate temperature, we will document some of the other benefits in terms of thermal management of the die under long term operation, reliability effects that arise from rise of chip junction temperature, and lastly a discussion of the relevant cooling technologies that have been introduced by technologists to implement the localized and globalized intra-die and inter-die cooling solutions. The following enumerations have been adapted from the classic reference article of Professor Kaustav Banerjee et al. from their published article, "Cool Chips: Opportunities and Implications for Power and Thermal Management" [39]. Cooling of MOSFET to reduced substrate temperature operation below room temperature benefits the back-end-of-line (BEOL) performance and reliability. Lower operating temperatures lead to smaller wire resistance per unit length, noting that metals exhibit a positive temperature coefficient and semiconductors a negative temperature coefficient. The reduction in wire resistance further enables delay in signal lines and static IR drop—a cause of major concern in extended power/ground rail-to-rail voltage application. Reliability of interconnect lines have been mostly associated with electromigration (EM) tolerance and inter-layer dielectric breakdown, also known as Time Dependent Dielectric Breakdown (TDDB), and these two metrics are found to be adequately devised by the engineering of chip cooling. This is due to the fact that at reduced thermal energy of electrons, the electrons have reduced thermionic drift velocity and also due to the increase of mean free path between collisions at lowered carrier or substrate temperature, the electrons therefore experience a lower statistical number of collisions stemming the otherwise observed growth of localized concentrations induced hot spots and steep thermal gradients, both of which are the known precursors to electromigration (EM) induced failure. TDDB is caused by defect induced conductive bridging of dielectric that exists between neighboring metal lines and also as a gate dielectric that separates the gate to channel conduction when the transistor is in the OFF state or the gate current is meant to be negligibly small. During the time the transistor is biased in the ON state, electrons and holes as generated by impact ionization near the drain can surmount the oxide tunneling barrier near the drain and create oxide traps and defects in the oxide, essentially serving the catalysts for TDDB failure at a later stage, by essentially forming the conductive bridging filament by oxide traps accumulation. At reduced temperature, the thermal energy of the electrons and holes are considerably low and at a given drain and gate bias, statistically a reduced number of electrons and holes are injected into the oxide, creating a substantially reduced number and distribution of traps along the oxide energy band, and hence TDDB lifetime of gate dielectric or inter-layer dielectric becomes more extended as reliability measures. The benefits of lower temperature operation in the BEOL wire scaling is such that for semiglobal and global wires, more aggressive interconnect scaling (narrower width conductors with fixed or narrower spacing between them) can be allowed under cooled operation without degrading RC delay and EM reliability. On the other hand, this provision of scaling wires can offset any inductive effects that may become prominent due to reduced resistance per unit length at lower substrate temperature. Also at lower temperatures, intra-wire capacitance per unit length can be reduced significantly for smaller aspect ratio wires (reduced metal thickness with constant width) while maintaining the same resistance per unit length. This will lower the delay per unit length thereby enhancing the rate at which bits can be transmitted per unit chip frequency, i.e., bandwidth. Cooling techniques can be broadly classified into two types based on cooling power consumption. Passive cooling denotes cooling by conduction (heat sink) and or natural convection by air, while active cooling represents different types of cooling schemes with associated external cooling power. Typically, pure passive cooling is only applicable for systems with low power consumption due to its low heat removal capacity which is limited by the structural profile of the heat sink (size, number of fins, fins aspect ratio and spacing, thermal conductivity etc.) and surrounding temperature. Cooling techniques for sub-ambient temperature operation, including refrigeration and cryogenics, are only applicable for specialized use to achieve required performance when cost and cooling power consumption are not the primary concerns. The efficiency of various refrigeration techniques can be compared by the coefficient of performance (COP) which is determined by the ration of cooling capacity to power consumption by the refrigerator ($COP = Q_{cooling}/W_{power}$), where $Q_{cooling}$ is the cooling capacity and W_{power} is the power consumption by the refrigerator. Moreover, the limit of heat removal capacity (Q) can be further improved by using liquid refrigerants as coolants. The concept of microchannel cooling with liquid has been introduced and investigated for applications with higher heat removal requirements (large size or array of high performance ICs). The latest technological development in properly designing microfabrication based cooling devices ensures the design of innovative microcoolers such as solid state thin film thermoelectric coolers (TEC) that will deal with the reliability concerns that emanate from the presence of hot spots (larger thermal gradient) on the substrate of high performance ICs. Recently, thin film TEC is becoming an attractive option mainly due to its compact structure and larger cooling capability [40, 71].

In this introductory review and chronology, the authors have summarized the significant potential benefit of low substrate temperature on scaled device performance both from leakage power concern and drive current enhancement. Additionally the scaling induced bottlenecks that vie with realizing the reduced temperature operation of MOSFET have been systematically addressed, and possible remedies based on device parametric engineering have been identified. The authors also briefly discussed the various cooling device solutions that have been developed by the microelectronic professionals to implement at the device manufacturing level. The next chapters onwards will concentrate on modeling solutions from the perspective of lower substrate temperature operation on threshold voltage control and its variability assessment and comparing the threshold voltage distribution profile in typical 100K–500K substrate temperatures distribution for various technology nodes considering mostly the long channel n-MOSFET case and projecting the tolerable threshold voltage window down to 10 nm technology node by judicious analysis of data from simulations performed on long channel n-MOSFET counterparts.

CHAPTER 2

Step-by-Step Computation of Threshold Voltage as a Function of Substrate Temperatures

2.1 ESSENTIAL MODELING EQUATIONS FOR COMPUTATION OF THRESHOLD VOLTAGE OF N-CHANNEL MOSFET AS A FUNCTION OF SUBSTRATE/LATTICE TEMPERATURE

In this chapter, we enunciate the modeling equations in sequence that can be accurately used to compute threshold voltage V_{th} of n-channel MOSFET as a function of substrate or lattice temperature. Chapter 1 stressed the importance of threshold voltage control from the perspective of scaling in order to enhance both on-current performance and minimize off-state leakage current. Chapter 1 also underscored the important observation that when the substrate temperature is reduced below room temperature, threshold voltage is increased, clearing the way for lower leakage current maintenance on the die chip. But an increase of threshold voltage must not be misconstrued as loss of on-current performance as channel mobility enhancement at lower substrate temperature will supersede the loss of drain current overdrive factor, i.e., ($V_{gs} - V_{th}$), and hence work as maintaining a highly superior I_{on}/I_{off} . The authors find it compelling that properly modeled equations, that when in sequence generate the threshold voltage of a long channel n-MOSFET, are not accurately described in many popular textbooks. One key misconception in unreliably computed values of threshold voltage even at room temperature is the generally accepted value of intrinsic carrier concentration n_i to be listed as $1.5 \times 10^{10} \mathrm{cm}^{-3}$ at $T = 300K$, and this value is not even close to the n_i value derived in the classic textbook by Robert F. Pierret [1] or experimentally determined value mentioned in classic textbooks by S. M. Sze et al. [5–7]. Absolutely precise calculation of n_i is of utmost importance, starting at room temperature, as the semiconductor bulk potential, which is a significant contributor to threshold voltage calculation, depends on correctly calculated n_i value. When the n-channel MOSFET is sufficiently long, threshold voltage V_{th} is high and on the order of 1 V or more. So if n_i is not correctly computed by proper modeling equations, the margin of error that will propagate in V_{th} calculation will be minimal. Now taking the aspect of device scaling and subjecting the MOSFET to enter short channel regime, V_{th} value, is substantially reduced from its near 1 V value and any error in n_i calculation will reflect in the bulk potential value and this will add significant uncertainty or spread in the nominal scaled value of V_{th}. In fact, with n_i value not properly modeled as a function of temperature, any kind of projection from long channel V_{th} value to 10 nm node with the V_{th} variability window definition will be largely error prone. Therefore, in this chapter the authors have underlined the key modeling equations to properly calculate n_i as a function of temperature, which can be used in conjunction with standard textbooks to solve some basic long channel MOSFET device theory based questions, albeit in a more accurate manner, taking the dispersion relationship of effective mass of electron and hole and band-gap modeling with temperature into account from cited literatures [15–18]. The authors used the most cited modeling approach for band gap as a function of temperature in [17] although more intensive and accurate analysis can be deduced from [18]. In reference to the textbook by Robert F. Pierret [1], the equation for intrinsic carrier concentration is,

$$n_i = (2.510 \times 10^{19}) \left(\frac{m_n^* \, m_p^*}{m_o \, m_o}\right)^{3/4} \left(\frac{T}{300}\right)^{3/2} e^{-\frac{(E_G - E_{ex})}{2kT}}. \tag{2.1}$$

In order to calculate n_i from the above equation, we would also need the temperature dispersion relationship of m_n^* and m_p^* and the temperature relationship of the band gap E_G. The value of E_{ex}, an exciton correction factor is indispensable for properly defining the n_i value at room temperature that is consistent with the modeling equations described in the papers by Barber et al. [15] and Kristin De Meyer et al. [16]. The modeling equations for m_n^* and m_p^* and E_G are illustrated in Equations (2.2)–(2.4).

$$\frac{m_n^*}{m_o} = 1.028 + \left(6.11 \times 10^{-4}\right) T - \left(3.09 \times 10^{-7}\right) T^2 \tag{2.2}$$

$$\frac{m_p^*}{m_o} = 0.610 + \left(7.83 \times 10^{-4}\right) T - \left(4.46 \times 10^{-7}\right) T^2. \tag{2.3}$$

Equations (2.2)–(2.3) reveal that m_p^* varies more strongly with temperature compared to m_n^*, also suggestive of different nonparabolicity factor dependence on temperature for conduction band E-k (Energy-wave number) profile and valence band E-k profile. The modeling equations for m_n^* and m_p^* are perfectly calculated for temperatures up to $T = 200$K. Beyond $T = 200$K, as the temperature is more reduced, the dispersion relationship has no proven experimental validation, hence using these equations for calculation of n_i at $T = 100$K might generate a slight error in $\left(\frac{m_n^* \, m_p^*}{m_o \, m_o}\right)^{3/4}$ calculation, but the authors (from ultimate simulations of threshold voltage versus temperature, where the temperature down to 100K has been taken into account) found no imperfect correlation from the plot of $V_{th} - T$ extracted through these modeling equations. Hence, Equations (2.2) and (2.3) have been used down to $T = 100$K to calculate n_i from (2.1).

The band gap $E_G - T$ modeling can be expressed by the following equation,

$$E_G (T) = E_G (0) - \frac{\alpha T^2}{(T + \beta)}, \tag{2.4}$$

where α and β are constants chosen to obtain the best fit to experimental data. The values of $E_G(0)$, α and β are given by,

$$E_G(0) = 1.170 \text{ eV}$$
$$\alpha = 4.730 \times 10^{-4} \text{ eV/K}$$
$$\beta = 636 \text{ K}$$

where T is in Kelvin.

Another important parameter that needs to be properly modeled as a function of temperature is intrinsic Fermi energy level of silicon E_i in eV and again taking reference to [1, 2],

$$E_i(T) = \frac{E_c(T) + E_v(T)}{2} + \frac{3}{4}kT \ \ln\left(\frac{m_p^*(T)}{m_n^*(T)}\right). \tag{2.5}$$

In Equation (2.5), being reference level, $E_V(T)$ is made equal to 0 eV resulting in $E_c(T) = E_G(T)$. Now Equations (2.2), (2.3), and (2.4) can be used to calculate E_i in eV at any T.

Next we need to properly model the equilibrium Fermi energy in the bulk E_F for p-type silicon substrate. It is expressed as,

$$E_F(T) = E_i(T) - kT \ \ln\left(\frac{p(T)}{n_i(T)}\right). \tag{2.6}$$

In Equation (2.6), $p(T)$ is only extrinsic near room temperature $T = 300$K. At $T = 200$K or below $p(T)$ is accurately defined by incomplete ionization term and at substantially elevated temperature than $T = 300$K, for example $T = 400$K to 600K, $p(T)$ is defined by a modified equation that stresses the incorporation of exponentially growing n_i term with T, and $p(T)$ is enhanced from its room temperature doping value to near intrinsic concentration value at these temperatures. The modeling of $p(T)$ as a function of T from very low temperature to high temperature is more important, as the doping of p-type substrate is low, i.e., between 10^{14}/cm^3 to 10^{16}/cm^3, as both the incomplete ionization effect contributions at lower T and intrinsic carrier concentration at higher T significantly alter the final calculated value of $p(T)$ for this mentioned doping values range, and hence $E_F(T)$ from Equation (2.6) as we simulate it versus all temperature ranges from 100–500K.

The computation of $p(T)$ involves the three distinct regions of $p(T)$ device physical attributes at different temperatures. At lower T, approximately from $T = 100$K to $T = 250$K, $p(T)$ is computed utilizing the concept of incomplete ionization of dopants. Near $T = 300$K to approximately $T = 400$K, $p(T)$ is computed using extrinsic region of dopant criteria and above $T = 400$K to $T = 600$K, $p(T)$ has a modified extrinsic modeling equation which incorporates the effect of exponentially growing value of n_i as T increases. The effect of n_i makes the $p(T)$ term larger than its typical extrinsic region value and at very high substrate temperature, n_i term dominates substantially to make the $p(T)$ value be defined by intrinsic condition rather than extrinsic condition. For p-type substrate, we first discuss the incomplete ionization region for acceptors N_A.

$$p = \frac{p_\mu}{2}\left[\left(1 + \frac{4N_A}{p_\mu}\right)^{0.5} - 1\right] \tag{2.7}$$

$$p_\mu = \frac{N_V}{g_A}e^{\frac{-(E_A - E_V)}{kT}} \tag{2.8}$$

$$N_V(T) = 2\left(\frac{2\pi m_p^*(T)\,kT}{h^2}\right)^{3/2}. \tag{2.9}$$

First by using Equation (2.9), $N_V(T)$ as a function of T in the range 100–200K is computed using the dispersion relationship of $m_p^*(T)$. Then using the generally accepted empirical constants g_A and E_A to be 4 and 0.045 eV for boron acceptors, $p_\mu(T)$ is calculated from (2.8). Finally, $p(T)$ is calculated from (2.7) using p_μ and the full ionization value of N_A. As it has been mentioned before, proper modeling of incomplete ionization at carrier freeze out zone, i.e., $T = 100$K to $T = 200$K, becomes necessary as the substrate doping is low, typically in the range of 10^{14}–$10^{16}/\text{cm}^{-3}$, and as has been mentioned in the introductory chapter, low temperature MOSFET operation employs the device doping near its intrinsic value to accrue the advantage of higher carrier mobility, and hence the modeling equations quoted here will be of greater importance from the aspect of low to moderate substrate doping. The extrinsic and intrinsic region binning of the $p(T)$ can be lumped into one integrated equation given by,

$$p(T) = \frac{N_A}{2} + \sqrt{\left(\frac{N_A}{2}\right)^2 + n_i(T)^2}. \tag{2.10}$$

Using (2.7)–(2.10), one continuous $p(T)$ profile can be generated for different T regions, i.e., freeze out region (low T), extrinsic region (moderate T), and intrinsic region (higher T).

At this stage, we can embark on developing the MOSFET threshold voltage equation in the long-channel regime with no consideration of short channel effects and including the effect of substrate temperature. As has been found in standard textbooks, the computation of φ_{ms} or metal-to-semiconductor work function difference,

$$\varphi_{ms}(T) = \frac{1}{q}\left[\varphi_M - \chi - (E_c - E_F(T))_{bulk}\right]. \tag{2.11}$$

The parameter $(E_c - E_F(T))_{bulk}$ provides the key temperature dependence effect to ϕ_{ms} otherwise excluded in standard textbooks. Now $(E_c - E_F(T))_{bulk}$ for p-type substrate is defined as,

$$E_C - E_F(T) = (E_G(T) - E_i(T)) + kT\,\ln\left(\frac{p(T)}{n_i(T)}\right). \tag{2.12}$$

For aluminum gate material, $\varphi_M = 4.28$ eV and for silicon $\chi = 4.01$ eV.

The near accurate value of φ_B is given by,

$$\varphi_B = KT \ln\left(\frac{p(T)}{n_i(T)}\right), \qquad (2.13)$$

where $p(T)$ is defined from incomplete ionization (lower T) to fully ionized (near room temperature) to intrinsic region (higher T).

The more accurate value of φ_B can be calculated in the following equation,

$$\varphi_B = E_i(T) - E_F(T). \qquad (2.14)$$

Now we embark on modeling the one-dimensional long channel regime threshold voltage as a function of substrate temperature,

$$V_T = \varphi_{ms}(T) + 2\varphi_B + \frac{\sqrt{2q\varepsilon_s\varepsilon_o p(T)2\varphi_B}}{C_{ox}}. \qquad (2.15)$$

The term in (2.13), C_{ox} is the gate capacitance per unit gate area (F/cm^2) and is given by,

$$C_{ox} = \frac{\varepsilon_{ox}\varepsilon_o}{t_{ox}}, \qquad (2.16)$$

where t_{ox} is the gate oxide thickness.

The source to body potential V_{sb} is a very effective metric to control on-chip and die-to-die threshold voltage variation and also to effectively control the chip subthreshold leakage current and chip on-current. The substrate bias effect can be modeled and integrated to Equation (2.15),

$$V_T = \varphi_{ms}(T) + 2\varphi_B + \frac{\sqrt{2q\varepsilon_s\varepsilon_o p(T)(2\varphi_B + V_{sb})}}{C_{ox}}. \qquad (2.17)$$

In Equation (2.15), V_{sb} is > 0 for negative substrate bias and V_{sb} is <0 for positive substrate bias. In all the above V_T modeling equations, the source potential V_s is ideally set to zero and acts as a reference point for both source to gate voltage V_{gs} and drain to source voltage V_{ds} .

CHAPTER 3

Simulation Outcomes For Profile of Threshold Voltage As a Function of Substrate Temperature Based on Key Device-Centric Parameters

3.1 SIMULATION OUTCOMES OF VARIOUS n-MOSFET DEVICE PARAMETERS INCLUDING THRESHOLD VOLTAGE AS A FUNCTION OF TEMPERATURE

In Chapter 2, we systematically outlined the analytical modeling equations for long channel conventional MOSFET to define various device parameters that are an integral part of threshold voltage computation and also derived the temperature dependence equations for these parameters. In this chapter, we will chronologically present the modeling outcome plots of these device parameters and their variations with substrate temperature followed by essential threshold voltage as a function of substrate temperature as the significant modeling outcome for this chapter. This material is appropriate for undergraduate and graduate level MOSFET modeling course material with staple concentration on long channel MOSFET based threshold voltage modeling as a function of substrate temperature where all device centric parameters that define the threshold voltage of a MOSFET have been precisely modeled by development of analytical equations which have not been completely analyzed or simulated in earlier available textbooks, conference papers, or journal editions.

3.2 SIMULATION OUTCOME OF INTRINSIC CARRIER CONCENTRATION (n_i) AS A FUNCTION OF SUBSTRATE OR LATTICE TEMPERATURE

As has been described in Chapter 2, intrinsic carrier concentration (n_i) is one of the fundamental building materials for threshold voltage. The bulk potential ϕ_B term depends on n_i by the logarithmic term, and imprecise calculation of n_i can significantly alter the actual analytical model predicted threshold voltage V_T value by either underestimating or overestimating the value of ϕ_B. This problem can be manifest and conspicuous even at room temperature $T = 300$K. One of the highly quoted and referenced textbook [10] states that n_i is of the value 1.5×10^{10} cm^{-3} at $T = 300$K and even if the experimentally deduced value of n_i is different than this quoted value it would add minor change in value of threshold voltage V_T. The comment the author made in [10] is true only when the long channel MOSFET device is considered where the computed threshold voltage is in the range of 1.2 V–1.8 V. As the channel length is scaled below 1 micron and of the order of 0.7 micron, the threshold voltage is already below 1 V and the error in precise calculation of n_i will be manifest through the incorrect value of ϕ_B term and therefore a value of n_i being equal to 1.5×10^{10} cm^{-3}, where precise analytical model equation for n_i calculation has not been developed, will result in erroneous value of ϕ_B and hence threshold voltage V_T. This scenario gets even abysmal as the channel length of n-MOSFET truly enters the short channel regime 0.35 micron or less, and the gross error in V_T will be cumulatively engendered as a proper analytical model based equation for n_i has not been adequately addressed and its value remains fixed at 1.5×10^{10} cm^{-3}. Now considering the current scenario of nanoscale MOSFET with nanowire MOSFET being in the projection, the threshold voltage V_T is of the order of 0.1 V or less and we will require almost 100% accuracy in ϕ_B term and hence V_T value so that the threshold voltage variability window is optimum and stringent to be within the bound of voltage scaling. In this scaling limit, a true and dependable value of n_i, even at $T = 300$K for silicon, is indispensable, and a modeling equation must be derived that calculates the reference value of n_i at $T = 300$K all the way from a long channel n-MOSFET perspective so that same reference value can be used for scaled MOSFET V_T calculation. The aim here is to model this n_i value as a function of temperature so that at least in the range of 200K–500K, the accuracy in calculation of n_i will generate accuracy in calculation of threshold voltage V_T. The plot generated here in this connection demonstrates the profile of n_i as a function of lattice/substrate temperature from 100K to 500K. The authors of this E-book have earlier commented that near $T = 150$K or below calculation of n_i will not be 100% accurate as the dispersion relationship of $m_p^*(T)$ and $m_n^*(T)$ are not accurately described at a temperature below $T = 150$K as befitting experimental values. But the n_i values demonstrated in this plot Figure 3.1 are accurately computed in the temperature range of $T = 200$K–$T = 500$K and the benefits of a low temperature operation can still be substantial for conventional MOSFET to present generation nanowire MOSFET even when the room temperature operation is lowered to a relatively modest 200K operation as the device engineers will have in their armory other device parameters centric architectural deployment to maintain superior high density integrated circuits and systems performance.

Figure 3.1 shows that at the lower temperature $T = 150$K, the value of n_i has been reduced substantially to affect the value of bulk potential, i.e., increasing its value and effectively increasing the value of V_T to the extent that the maximum tolerable value of V_T in the V_T variability window for ultra scaled MOSFET beyond 10 nm may be tripped. One significant aspect that can be utilized to control V_T value in Equation (2.13) is by selective ion implantation of dopants in the channel region closer to the source terminal. By selective phosphorous dose by ion implantation in

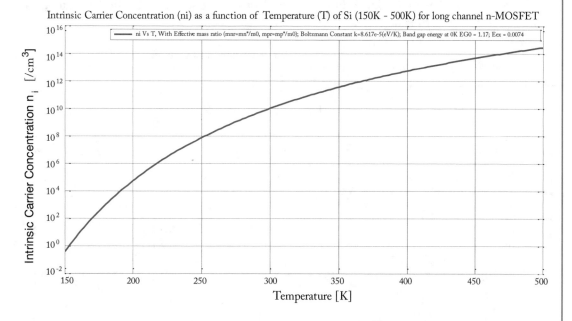

Intrinsic Carrier Concentration (ni) as a function of Temperature (T) of Si (150K - 500K) for long channel n-MOSFET

Figure 3.1: Intrinsic Carrier Concentration (n_i) of silicon as a function of temperature from $T = 150K$ to $T = 500K$.

the channel near the source, we can add a dose induced term $\frac{qD_i}{C_{ox}}$ where $qD_i < 0$ for phosphorous dopants and in this way we can maintain the scalability requirement of V_T for scaled voltage requirement when the substrate temperature is considerably below the room temperature. Now we proceed to demonstrate a key ingredient of modeling threshold voltage at lower than room temperature called incomplete ionization.

3.3   SIMULATION OUTCOME OF INCOMPLETE IONIZATION OF DOPANTS RELEVANT FOR LOWER SUBSTRATE TEMPERATURE OPERATION

Figure 3.2 demonstrates the dopant activation values from low temperature to higher than room temperature. At room temperature and higher the Fermi energy level E_F (eV) lies below the dopant energy level E_D (eV) so that all dopant energy levels are empty and conduction energy bands are full of dopant carriers. At substantially reduced temperature than room temperature, Fermi energy level E_F moves closer to conduction band and at extremely reduced temperature even inside the conduction band, and now the dopant energy level E_D is below the conduction band and as a result, some of the energy levels of E_D are occupied by the dopants and the conduction band is thus not populated by full 100% of the dopants that are added to the silicon bulk. The impact of incomplete ionization at lower temperature is more severe when the substrate dopants are higher or not in the range of $10^{14} - 10^{16}/cm^3$.

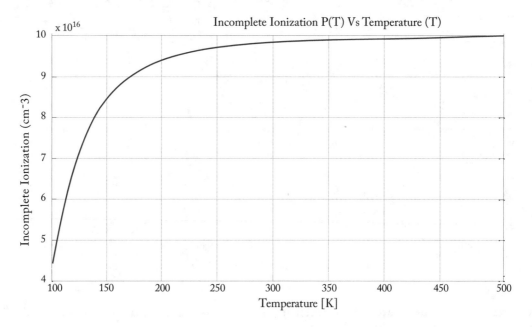

Figure 3.2: Incomplete ionization effect on substrate doping optimized for 1 micron channel length MOSFET from $T = 100K$ to $T = 500K$.

As the dopants of today's scaled MOSFETs are of the order of 10^{18} cm^{-3}, the Fermi energy level at the bulk will be driven closer to conduction band and the dopant energy level will turn out to be below E_F level, where the $E_D - E_F$ will be more negative at lower operational temperature and for the case of higher substrate doping. This will introduce incomplete ionization much earlier in lower T than the case when the substrate doping was in the moderate range. Also at the lower operational temperature, the incomplete ionization effect will result in more reduced active ionized carriers for higher substrate doping required to sustain scaling than for the case when substrate doping is moderate or low.

3.4   SIMULATION OUTCOME OF FERMI ENERGY LEVEL E_F (EV) AS A FUNCTION OF TEMPERATURE

At lower substrate temperature the Fermi energy level for p-type silicon substrate moves closer to valence band energy level and almost coincides with it at substantially reduced temperature.

As temperature is increased to much above room temperature, the Fermi energy level E_F in the beginning rises fast and then gradually reaches closer to the intrinsic Fermi energy E_i value which after incorporation of the exact Equation (2.5) is lower than the originally referred value as $\frac{E_C + E_V}{2}$ eV. Figure 3.3 is depicted below.

Fermi energy (EF) as a function of Temperature (T) range 100K–500K for long channel n-MOSFET

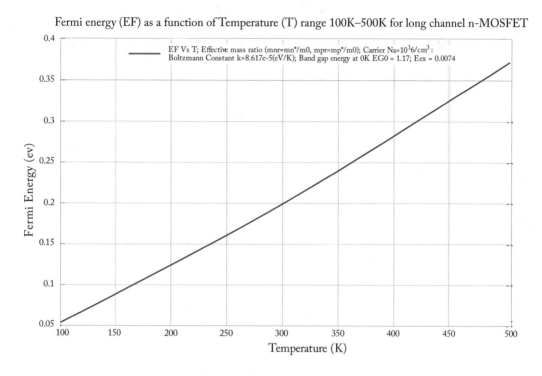

Figure 3.3: Fermi energy level E_F with respect to the valence band E_V for the reference doping for 1 micron channel length n-MOSFET from $T = 100$K to $T = 500$K.

3.5   TEMPERATURE DEPENDENCE OF FLAT BAND VOLTAGE ϕ_{ms} (V)

Considering the metal Fermi level is not affected by substrate temperature change but remains fixed at its $T = 300$K, the flat band voltage which is essentially the metal-to-semiconductor work function difference is most negative for Aluminum metal gate and p-type substrate Fermi energy level at $T = 100$K and gradually reduces as T is increased to 500K. This is due to the fact that both the bulk potential term ϕ_B and $E_G/2$ are larger at substantially lowered substrate temperature value, making semiconductor work function ϕ_s at the semiconductor bulk considerably higher, making the ϕ_{ms} term more negative. As temperature of the substrate is increased to room temperature and beyond, both ϕ_B and $E_G/2$ get reduced and as a result ϕ_{ms} becomes less negative. There is an added beneficial effect at lowered substrate temperature operation where ϕ_{ms} is more negative. It will in effect neutralize the impact of reduced n_i at lower T on threshold voltage V_T which is to up shift the threshold voltage through increased value of the bulk potential term. The more negative ϕ_{ms} term will still control the extra shift on V_T that arises through reduced temperature operation and will maintain the required scaling based V_T variability. Figure 3.4 shows the flat band voltage dependence on temperature is given in the space below.

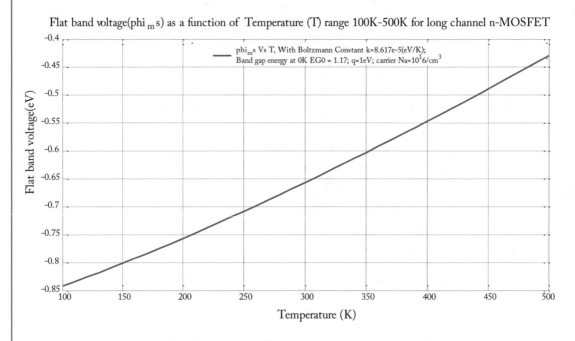

Figure 3.4: Flat Band Voltage dependence of n-channel MOSFET for channel length 1 micron and optimized substrate doping as a function of substrate temperature from $T = 100K$ to 500K.

3.6 P-TYPE SUBSTRATE n-CHANNEL MOSFET BULK POTENTIAL DEPENDENCE ON SUBSTRATE/LATTICE TEMPERATURE

Figure 3.5 shows the bulk potential ϕ_B dependence on substrate temperature. At very substrate temperature near 100K or below, the Fermi level E_F for p-type substrate n-channel MOSFET is almost pinned to valence band level E_V, whereas intrinsic Fermi energy level E_i is at $E_G/2$ where E_G has increased from its room temperature value. As a result the difference $(E_i - E_F)$, which is ideally the ϕ_B, increases substantially at lower T. As substrate temperature is increased to achieve room temperature value and beyond, the bulk potential ϕ_B is gradually reduced as E_F reaches closer to E_i. At much higher substrate temperature, E_i goes below the term $\frac{E_c + E_V}{2}$ through the subtractive effective mass dependence and at the same time E_F is also going closer to E_i. Hence, the difference $(E_i - E_F)$ decreases at a much faster rate if correct numerical simulation and modeling equation are employed rather than dopant based logarithmic dependence Equation (2.13). This will make the bulk potential to undergo a steeper descending slope as temperature is increased further in the range of $T = 500\text{K}$ or higher.

Bulk potential (phi$_b$) as a function of Temperature (T) range 100k–500K for long channel n-MOSFET

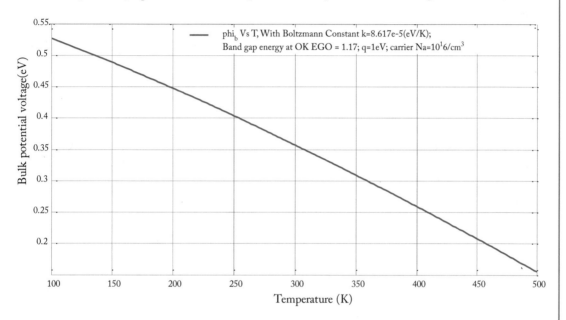

Figure 3.5: Bulk potential dependence on substrate temperature for p-substrate 1 micron channel length n-MOSFET with optimized substrate doping.

3.7 DEPENDENCE OF THRESHOLD VOLTAGE V_T OF n-CHANNEL MOSFET ON SUBSTRATE TEMPERATURE FOR 1 MICRO CHANNEL LENGTH MOSFET

It is known to the semiconductor device modeling professionals and community that long channel MOSFET threshold voltage follows linear inverse temperature dependence. Nevertheless, previous publications have not presented the modeling equations and show by simulation outcome that this is the case. The authors of this textbook in this context have identified the true V_T defining parameters and their associated modeling by temperature dependence and, as the following figure will reveal, a near perfect inversely linear temperature dependence has been underscored for threshold voltage of long channel conventional MOSFET. As the readers are expected to closely observe the $V_T - T$ profile, the V_T in fact bends back from linear dependence closer to $T = 100K$ and sharply bends down at $T = 500K$. This is due to the incorporation of incomplete ionization effect at substantially reduced substrate temperature closer to $T = 100K$, which reduces the contribution from $p(T)$, which is not fully ionized to reference substrate doping level. At $T = 500K$ or higher, V_T bends down from true linear dependence as the ϕ_B term decreases at a faster rate than its more generic logarithmic dependence on doping value, mostly quoted in literature. We now express this in Figure 3.6.

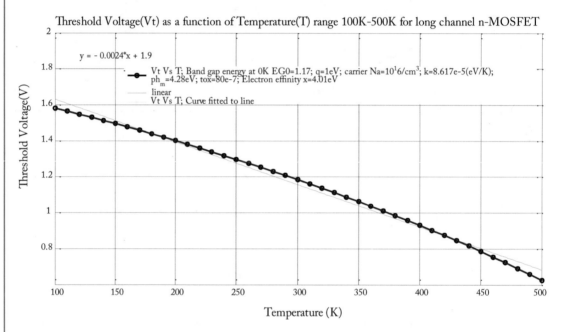

Figure 3.6: Threshold voltage dependence on substrate temperature from $T = 100K$ to $T = 500K$ for conventional long channel 1 micron channel length n-MOSFET with optimized substrate doping.

3.7.1 MODELING IMPACT OF INCOMPLETE IONIZATION ON THRESHOLD VOLTAGE AT THE FREEZE-OUT TEMPERATURE REGION: A CLOSER LOOK

A closer look at the proper requirement of modeling of incomplete ionization effect on threshold voltage variation at the vicinity of freeze-out temperature region reveals important simulation outcomes for different substrate doping values. For a scaled MOSFET, as the required substrate doping increases, the following analysis will reveal that the impact of incomplete ionization of acceptors becomes more pronounced and the threshold voltage will not rise as linearly as would be predicted, simple room temperature threshold voltage modeling expression extending down to temperatures around 100K. On one hand this proves that as the required substrate doping is high for aggressively scaled MOSFET and henceforth due to incomplete ionization, threshold voltage does not rise as much as would be predicted by linear profile, the impact on leakage current control will be aggravated as, due to lesser threshold voltage value than optimally expected, scaled MOS-FET will be prone to some marginal leakage current increase, as reduction of threshold voltage in the vicinity of 100K manifests itself through increasing impact of freeze-out at lower substrate temperature for higher substrate doping conditions. There is one beneficial impact that we see due to threshold voltage rise less than linear proportion from room temperature reference value. Due to less rise of threshold voltage, owing to freeze out effect at lower substrate temperature for higher substrate doping, the actual threshold voltage value will be amenably compatible with supply voltage reduction, and the threshold voltage variation window will be conveniently within the controlled supply voltage scaling. Now for analysis of this important freeze-out effect due to varied substrate doping we refer the readers to the following tabular data:

Table 3.1: Impact of freeze-out effect on carrier ionization for temperatures at 100K, 200K, and 300K

Na = 10^16	
p = (pmu./2).*(sqrt(1+((4.*Na)./pmu))-1)	
(TK)	P(T)
100	4.51E + 15
200	9.45E + 15
300	9.88E + 15

Na = 10^17	
p = (pmu./2).*(sqrt(1+((4.*Na)./pmu))-1)	
(TK)	P(T)
100	1.75E + 16
200	6.99E + 16
300	8.98E + 16

Na = 10^18	
p = (pmu./2).*(sqrt(1+((4.*Na)./pmu))-1)	
(TK)	P(T)
100	5.90E + 16
200	3.30E + 17
300	5.79E + 17

From the above table we can see that for a p-type acceptor substrate doping of $10^{16}/cm^3$ which is the optimum doping at $T = 300K$ for the reference gate length of 1 micron n-MOSFET, the ratio of ionized acceptors at $T = 100K$ and $T = 300K$ is 0.456 or 45.6% is ionized at this substrate doping for $T = 100K$. As the substrate doping is increased to $10^{17}/cm^3$ to let's assume for a channel length which will be less than 1 micron, the ratio of ionized acceptors at $T = 100K$ to $T = 300K$ is 0.195 or 19.5% is ionized at this substrate doping. Further increasing the substrate doping to $10^{18}/cm^3$, which will be required when channel length is substantially reduced from its 1 micron reference value, the ratio turns out to be 0.102 or 10.2% is ionized. So we can see from this trend that as the device designers are increasing the substrate doping to combat channel length reduction induced short channel effects, such as punch through and drain field penetration to the source junction, the lower substrate temperature operation substantially reduces the available dopant carriers due to the dominant effect of incomplete ionization for higher substrate doping which does not result in proportionate increase of threshold voltage from its room temperature or $T = 300K$ reference value. The reason for this observation from the simple energy band perspective is that for a fixed acceptor doping condition, the Fermi energy level moves closer to the valence band for higher substrate doping concentration value and the Fermi level is situated in the middle between the fixed acceptor energy level and the valence band reference level. For higher substrate doping value the difference between acceptor energy level and Fermi level is higher, and acceptor energy level is higher than Fermi energy level, therefore they are more filled than empty (generating holes in the valence band). As the substrate doping concentration is reduced, the Fermi level shifts to an upper level from valence band energy level edge, and the difference between acceptor and Fermi energy levels are reduced, although the acceptor level can be still above the Fermi level or the Fermi level can be slightly above than acceptor energy level. Then there will be emptier acceptor sites contributing to significant carrier holes in the valence band.

3.8 THRESHOLD VOLTAGE DEPENDENCE ON SUBSTRATE TEMPERATURE FOR DIFFERENT SUBSTRATE DOPING CONDITIONS FOR AN n-CHANNEL MOSFET

Selection of proper substrate doping and optimizing it through ion-implanted profile engineering is one of the most important determinants of optimum threshold voltage value at a given technology node, preferably when the channel length is in the long channel regime. Here we present the threshold voltage dependence on substrate or lattice temperature from $T = 100K$ to $T = 500K$ for different substrate doping conditions, where we first concentrate the result on the optimum substrate doping chosen for a 1 micron n-channel MOSFET and then take some reduced values from this nominal value point and some higher values from this nominal value point. The following Figure 3.7 systematically shows the threshold voltage-substrate temperature plots for nominal substrate doping of $10^{16}/cm^3$ and then for $6.5 \times 10^{15}/cm^3$, $7.5 \times 10^{15}/cm^3$, $9 \times 10^{15}/cm^3$ on the lower extremes from optimal and then for $2.5 \times 10^{16}/cm^3$, and $4 \times 10^{16}/cm^3$ for higher extremes from optimal.

From Figure 3.7, we can see the deviations of threshold voltage (V_T) as a function of substrate temperature (°K) from its expected inverse linear relationship are more pronounced as the substrate doping is increased. Although at the mid temperature level the $V_T - T$ profile is more or less in line with the fitted linear function for each doping concentration case, the deviations are noteworthy at lower temperatures close to 100K due to the incomplete ionization effect and then rapid decrease from the linear decline curve at the higher temperature regions close to 500K and higher due to increasing intrinsic carrier generation and the reduced difference between intrinsic Fermi energy E_i and bulk Fermi level E_F when higher temperature effects are properly taken into account in the modeling equations. Both of these effects are stronger as the substrate

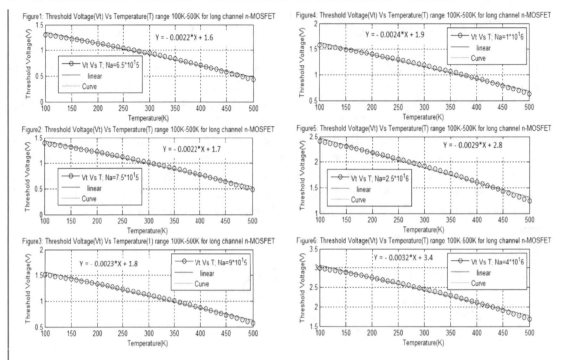

Figure 3.7: Dependence of threshold voltage value on substrate temperature for different substrate doping conditions.

dopant values are increased gradually from the lowest value chosen from the optimal condition to the highest value chosen above optimal condition. Therefore, Figure 3.7 can be used as a useful guideline for the device engineers to suitably choose the substrate doping to optimize V_T at lower substrate temperature because of the intrinsic performance benefits, taking into consideration the important incomplete ionization effect based nonlinearity that can handicap the achievable V_T at lower T if the substrate doping is less than optimized.

3.9   THRESHOLD VOLTAGE DEPENDENCE ON SUBSTRATE TEMPERATURE FOR DIFFERENT OXIDE THICKNESS FOR AN n-CHANNEL MOSFET

Gate oxide thickness (t_{ox}) is an important parameter that can be engineered by applying both device and process modeling optimization to implement the required threshold voltages value at a given channel length L and width W for different die locations on a wafer. Since the gate oxide thickness determines the vertical electric field that originates from applied gate to source or channel potential, the gate integrity, i.e., gate's ability to control channel's inversion charges at a given drain bias, is directly affected by the thickness of gate dielectric. As the gate length scales using the constant field criterion, gate oxide thickness scales at the same proportional rate and therefore at reduced channel dimensions, the gate's integrity to control channel inversion charges improves by dint of concomitant gate oxide thickness reduction. At reduced gate oxide thickness with the scaled voltage level both on gate and drain nodes, since the gate vertical field is higher for reduced gate oxide thickness, lateral drain field will not be able to penetrate significantly to the source since increased vertical electric field will compensate some drain field lines extending toward the channel region from the drain contact. Therefore, drain induced barrier lowering (DIBL), an effect which is dominant in a weak inversion region, will be less if the oxide thickness is scaled to the desired proportion, which as per the above explanation will extenuate this important short channel effect. Decreased gate oxide thickness also improves channel on/off ease of controllability and results in improved subthreshold slope, which is an important device centric performance feature for superior scaled MOSFET performance monitoring. With the supply voltage reduction, reduced oxide thickness enables maintenance of the required inversion charge density, which otherwise would have been reduced at not so scaled oxide thickness condition due to the requirement of reduced gate voltage. Reduced gate oxide thickness will exacerbate the short channel effects induced reliability failures, such as gate tunneling current leakage at reduced gate oxide thickness approaching the direct tunneling current regime when the gate potential is 0, slightly positive or slightly negative. During the gate bias and drain bias that institutes the operating condition when the MOSFET delivers the high on-current, significant hot carrier effects can result in some thermal momentum enhanced carrier electrons and holes tunneling through the reduced gate oxide thickness and creating traps in the oxide that lead to device transconductance and mobility degradation.

Since reduced gate oxide thickness improves gate integrity to control channel inversion, threshold voltage V_T is more suitably reduced at a proportionate rate determined by Equation (2.17) and from a low substrate temperature to high temperature threshold voltage dependence perspective this is beneficial, as at the reduced substrate temperature closer to 100K where as V_T is enhanced to a small amount due to an incomplete ionization effect of the dopants explained earlier, this slight increase can be nullified by the systematic V_T reduction effect caused by t_{ox} reduction. At the higher substrate temperature, since V_T falls off at a slightly faster rate from the straight line decline due to reduced $(E_i - E_F)$ value, this V_T will be further reduced if reduced t_{ox} needs to be employed, and this additional reduction of V_T at the higher temperature end for the long channel MOSFET will push the design V_T further lowered at the scaled MOSFET and increase subthreshold leakage current. Therefore, the benefit of MOSFET scaling on V_T adjustment can be most effectively orchestrated between a substrate temperature of 100K and a little over 300K, shunning the design of higher substrate temperature values. Since this E-book mostly concentrates an harnessing the aspects of reduced substrate temperature operation rather than room temperature, we potentially accrue all the device benefits on V_T mainly at substrate temperature between $T = 100K$ to $300K$ (room temperature) for the nominal oxide thickness 80 nm chosen for $L = 1$ micron channel length and applying a variation around it.

The following Figure 3.8 shows the threshold voltage dependence on different oxide thickness chosen, taking 80 nm as reference for substrate temperatures 100K to 500K.

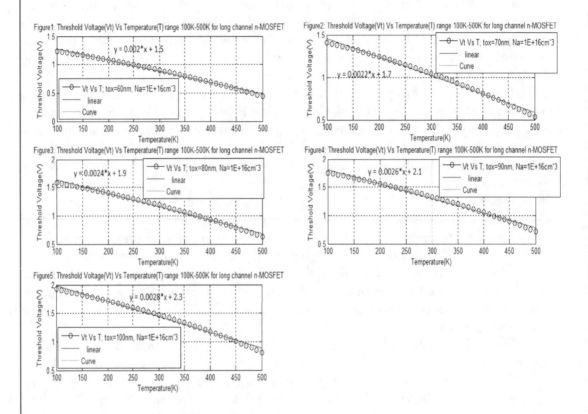

Figure 3.8: Threshold voltage dependence on oxide thickness for substrate temperatures from $T = 100K$ to $T = 500K$.

From Figure 3.8, it is evident that t_{ox} thickness control is a very potent factor, as we see the reduction or increase of gate oxide thickness from its nominal value does not affect the uniformity of V_T distribution compared to substrate dopant case of Figure 3.7 where higher substrate dopants, which need to be designed for scaled MOSFETs, cause significant non linearity in the V_T distribution profile at the temperature edges close to $T = 100K$ and $T = 500K$. Figure 3.8 also illustrates the conclusion that oxide thickness can be suitably chosen to maintain the uniformity of V_T distribution trend extracted from the reference oxide thickness condition over the applied substrate temperature ranges—a fact which will spur the device engineers who are concentrating on designing scaled n-MOSFET with maximum intrinsic and device centric performance boosts.

3.10 THRESHOLD VOLTAGE DEPENDENCE ON SUBSTRATE TEMPERATURE FOR NEGATIVE SUBSTRATE BIAS FOR AN n-CHANNEL MOSFET

When there is an urgent need in circuit operation to reduce the off-state leakage current or sub-threshold current flowing through devices integrated in a circuit, device designers employ negative substrate bias instead of more typically applied ground potential to shift the device V_T upward as is revealed by Equation (2.17) where a negative body potential and source being at ground makes $V_{SB} > 0$. For larger V_{SB} values, the depletion region widens from its substrate doping related maximum extension, and variation in V_T is small even when die to die dopant values can change by a small deviation from the reference doping. For smaller V_{SB} values and moderate to low V_T value, the depletion region width can vary to a measurable amount due to the dopant variation, due to process integration in die to die from its reference value, and this enhances V_T variation from die to die. A moderate to large V_{SB}, despite reducing die to die V_T variation, increases the drain to substrate reverse depletion width and channel potential, and for short channel length, the drain depletion width induced drain field can extend more toward the source and cause drain induced barrier lowering effect to enhance weak inversion related drain current increase at a substantially lower gate voltage. In order to comply with the voltage scaling and the imputable value on gate voltage so that a scaled V_T value condition can be optimized, the V_{SB} value thus needs to be ingeniously devised to be in accordance with supply voltage scaling. Since V_T values are already shifted upward for substrate temperatures less than 300K from its reference value at $T = 300$K, a minutely designed smaller positive V_{SB} value needs to be computed using modeling equations, as device designers do not wish to widen V_T variation window at the lower substrate temperature operating conditions. Overall, in order to comply with voltage scaling and ultra low voltage MOSFET operation, V_{SB} biasing must be afforded within the smallest possible design values window.

In Figure 3.9, we show the threshold voltage distribution profile as a function of substrate temperature from 100K to 500K at the reference doping and oxide thickness condition for 1 micron channel length for substrate biases of (i) $V_{SB} = 0.75$ V, (ii) $V_{SB} = 1.00$ V, (iii) $V_{SB} = 1.25$ V, and (iv) $V_{SB} = 1.5$ V. As in accordance with the explanation enumerated, lower V_{SB} values preserve the uniformity of V_T distribution along the substrate temperatures, particularly near 100K substrate temperature operating condition where a moderate to large V_{SB} value can cause non linearity in the profile by shifting V_T upward. This is also the region where an incomplete ionization effect can interact with the V_{SB} values to set up a V_T that is too much deviant from the linearly defined characteristic $V_T - T$ profile.

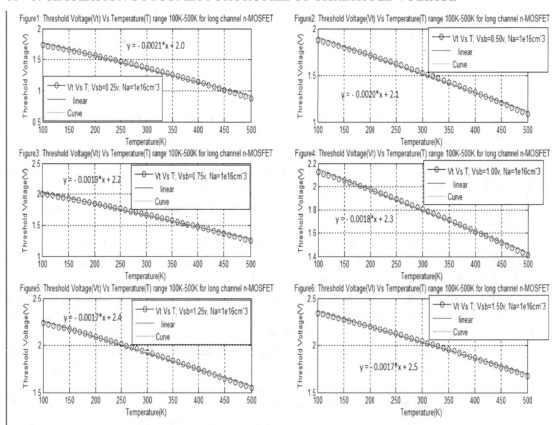

Figure 3.9: Threshold voltage variation as a function of substrate temperature from 100K to 500K for different negative substrate biases ($V_{SB} > 0$).

3.11 THRESHOLD VOLTAGE DEPENDENCE ON SUBSTRATE TEMPERATURE FOR POSITIVE SUBSTRATE BIAS FOR AN *n*-CHANNEL MOSFET

Although in most device applications from the leakage current control perspectives, a negative body biasing scheme is most prominent, occasionally we can bias the body positively to cause source to body potential $V_{SB} < 0$. As is evident from Equation (2.17), a negative V_{SB} reduces the nominal V_T value through the square root term—a contribution from the depletion charge. As the device die to die V_T value must be scaled and must remain small, complying generalized voltage scaling, negative V_{SB} is a very potent and efficient way to set the smaller V_T values as required in a chip in very large scale integration of dies. There is an important observation of how to set the V_{SB} potential, i.e., source at ground potential and body at a negative potential or source at negative potential and body at ground potential. When source is at ground potential, the true V_{GS} condition at the source junction is established which controls the barrier for carriers being injected into the channel from source side. As the p-substrate is biased positive in this case, the p-substrate to n-source junction is forward biased, and some of the electrons being injected into the channel will be attracted to the p-substrate potential as minority carriers diffusion, and this will result in fewer electrons being injected into the channel at small $V_{GS} > 0$ and also a small $V_{SB} < 0$. Hence, although the nominal V_T is in accordance with scaled MOSFET requirements, a definite amount of variation of electron density or number of electron carriers being injected from the source side will introduce a definite variation in V_T from die to die. Let's now investigate the other scenario. When the source potential is negative and body potential is ground, V_{GS} is enhanced even for small V_G value and this will lower the barrier additionally at the source side for electron injection into the channel, and a negative source also will create a repulsive field around electrons being injected into the channel so that they will not backscatter to the source. The p-substrate n-source is still forward biased but, electrons as minority carriers experience not so enhanced attractive potential at substrate contact which is at ground potential; rather it is the minority holes that diffuse along the junction on the way to being collected by the attractive negative source potential and recombine with source side injected electrons into the channel. Since hole mobility is lower than electron, the recombination effect will be not so crucial to change or alter the injected carrier density into the channel from the source side as has been observed with the previous bias condition for $V_{SB} < 0$. Therefore, although from ease of manufacturability and volume fabrication, substrate bias being set to positive value can be more effectively accomplished, we might look into the latter option being discussed where we still apply ground potential to substrate and instead bias the source to slightly negative value. One of the other important modeling aspects that evolves from $V_{SB} < 0$ bias condition is that in Equation (2.17): under the square root term in the numerator, the bulk potential is temperature dependent and slowly changes from high positive value at lower T to small positive value at higher T. Hence, we need to carefully apply the value on V_{SB} as a fixed large or even moderate negative V_{SB} value can make the overall value inside the square root negative, and this will result in V_T values becoming imaginary for moderate to higher substrate temperature regions where the bulk potential is gradually and then sharply decreasing. The obvious choice of V_{SB} value is then determined by the lowest value of bulk potential around $T = 500$K and V_{SB} must be lower than this bulk potential value at a reference doping condition. Figure 3.10 details the plot of threshold voltage as a function of substrate temperature for (i) $V_{SB} = -0.15$ V, (ii) $V_{SB} = -0.20$ V, and (iii) $V_{SB} = -0.30$ V. Again, just like positive substrate bias case, a smaller V_{SB} value for positive substrate bias will preserve the uniformity of V_T profile by maintaining small variations closer to $T = 100$K, dominated by an incomplete ionization effect and at $T = 500$K where V_T sharply declines from the linear curve. From close observation of Figure 3.9 and Figure 3.10, it can be concluded that compared to the negative substrate bias condition, the positive substrate bias condition imparts significant non linearity in V_T distribution profile even for very small V_{SB} values, and hence positive substrate bias and substrate temperature both will pose a trickier challenge for device designers to set their targeted V_T values.

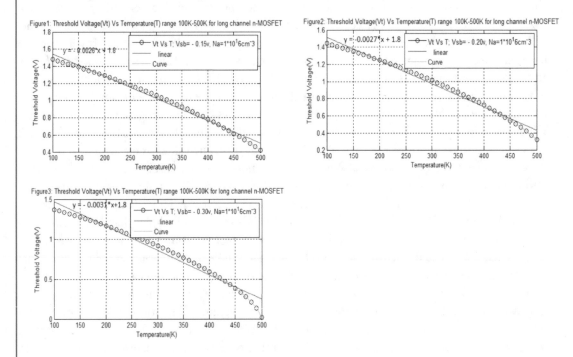

Figure 3.10: Threshold voltage variation as a function of substrate temperature for positive substrate bias potential.

CHAPTER 4

Scaling Projection of Long Channel Threshold Voltage Variability with Substrate Temperatures to Scaled Node

4.1 MODELING AND SIMULATION RESULTS FOR A LONG CHANNEL MOSFET AS CHANNEL LENGTH IS SCALED FURTHER

Since one of the aims in this book is to analyze and model the impact of channel length scaling on threshold voltage dependence on temperature, for the same normalized width where we have taken the width sufficiently wide to disregard narrow width effects, a comparative data centric analysis will be presented in this section for the reference n-channel MOSFET with $L = 1$ micron and oxide thickness $t_{ox} = 80$ nm and a scaled n-channel MOSFET with $L = 0.7$ micron and oxide thickness $t_{ox} = 56$ nm. The reference doping for 1 micron channel length device is $N_A = 10^{16}/cm^3$ and for 0.7 micron channel length $N_A = 1.5 \times 10^{16}/cm^3$. First we compare the effect of doping variation on the threshold voltage distribution profile in the temperature range 100–500K. Although the simulations included the continuous treatment of the entire temperature range between 100K to 500K, the table below shows the comparisons for specific temperature values that will serve as guidelines for setting the threshold voltage at the operation case of the circuit. One of the key factors to note here is the threshold voltage standard deviation which for these devices will be of the order of a few 100 mV.

From Table 4.1 we can see that threshold voltage standard deviation at $T = 300$K is around 0.297 mV and at lower substrate temperature below 300K the standard deviation increases from $T = 300$K in a rather non-uniform manner, which is indicative of the incomplete ionization effect and its varied impact on the different substrate doping values. At temperatures above 300K, we see a progressive reduction of standard deviation of threshold voltage. This is due to the fact that threshold voltage itself decreases as temperature is increased, and this plays a role in reduction of threshold voltage standard deviation alongside the general uniformity of full dopant ionization.

From Table 4.2, it is clearly evident that scaling improves the threshold voltage standard deviation as we can speculate that a scaling of $L = 1$ micron to $L = 10$ nm makes the standard

Table 4.1: V_T value for the selected dopant variations from mean value ($L = 1\,\mu$m)

Value	Na=6.5x10^15	Na=8x10^15	Na=10^16	Na=1.2x10^16	Na=2x10^16	STDEV
T=100K	1.3112	1.4354	1.5841	1.7186	2.1681	0.270596
T=150K	1.2326	1.3542	1.4999	1.6316	2.0722	0.324274
T=200K	1.1456	1.2641	1.4061	1.5345	1.9644	0.31625
T=250K	1.0508	1.1658	1.3035	1.4282	1.8459	0.307105
T=300K	0.9483	1.0593	1.1922	1.3127	1.7166	0.29677
T=350K	0.8375	0.944	1.0717	1.1874	1.5762	0.285359
T=400K	0.7172	0.8187	0.9406	1.0512	1.4234	0.272848
T=450K	0.585	0.6812	0.7968	0.9018	1.2561	0.259316
T=500K	0.4366	0.5271	0.6361	0.7352	1.0705	0.24499

Table 4.2: V_T value for the selected dopant variations from mean value ($L = 0.7\,\mu$m)

Value	Na=1x10^16	Na=1.2x10^16	Na=1.5*10^16	Na=1.80x10^16	Na=2x10^16	STDEV
T=100K	1.1734	1.268	1.3961	1.512	1.584	0.169043
T=150K	1.1041	1.197	1.3229	1.4368	1.5075	0.166093
T=200K	1.0274	1.1183	1.2414	1.3527	1.4219	0.162409
T=250K	0.944	1.0325	1.1523	1.2608	1.3281	0.158142
T=300K	0.854	0.9397	1.0559	1.1611	1.2265	0.153362
T=350K	0.7571	0.8398	0.9519	1.0535	1.1166	0.148014
T=400K	0.6526	0.7319	0.8395	0.9371	0.9977	0.142094
T=450K	0.5388	0.6144	0.7172	0.8104	0.8684	0.135712
T=500K	0.4129	0.4847	0.5822	0.6708	0.726	0.1289

deviation 10 times lower at $L = 10$ nm from its $L = 1$ micron counterpart. Again we observe from this data stream that at lower operational substrate temperatures less than 300K, standard deviation increases and at above room temperature, standard deviation decreases, for the same reasons mentioned before. Now we analyze the threshold voltage standard deviation for the two channel length prototypes for the impact of different gate oxides with different substrate temperatures. Since the gate thickness modulation controls the gate integrity of the channel inversion charge, its impact on threshold voltage introduces less non uniformity compared to variations of dopant density as fluctuations in surface band bending in addition to vertical field determined by oxide thickness have direct bearing on fluctuations in dopant density. Due to selection of a nominal dopant density but varied oxide thickness, the field induced surface potential bending has a one-to-one correlation with oxide thickness but the fluctuations in surface potential will be nonexistent due to selection of a single nominal dopant density value.

Table 4.3: V_T value for the selected oxide thickness variations from mean value ($L = 1\,\mu m$)

Value	tox=60nm	tox=70nm	tox=80nm	tox=90nm	tox=100nm	STDV
T=100K	1.2418	1.413	1.5841	1.7552	1.9264	0.270596
T=150K	1.1701	1.335	1.4999	1.6647	1.8296	0.260682
T=200K	1.0905	1.2483	1.4061	1.5638	1.7216	0.249456
T=250K	1.0039	1.1537	1.3035	1.4533	1.6031	0.236855
T=300K	0.9104	1.0513	1.1922	1.3332	1.4741	0.22283
T=350K	0.8095	0.9406	1.0717	1.2027	1.3338	0.20724
T=400K	0.7006	0.8206	0.9406	1.0606	1.1807	0.189768
T=450K	0.5818	0.6893	0.7968	0.9042	1.0118	0.169957
T=500K	0.4502	0.5431	0.6361	0.7291	0.8219	0.146951

From Table 4.3, it is clear that the nominal standard deviation of threshold voltage at $T = 300K$ is itself lower compared to the dopant distribution case. Also the standard deviation gradually increases to higher values as substrate temperature is lowered from room temperature. This clearly explains that at lower temperature, the vertical field lines that penetrate from gate potential to the ionized acceptors, which are sparsely distributed, are less intense, less direct, and less uniform, resulting in additional threshold voltage standard deviation. But this increase is rather uniformly transpired over the temperature range at the lower extremity as the substrate dopant density variation impact on surface potential fluctuations is absent in this case. As the substrate temperature is increased from room temperature, the standard deviation of threshold voltage taking considerations of variations in oxide thickness is rather uniformly decreased. This is due to the much lower threshold voltage values at these temperatures and also due to the more even distribution of ionized acceptors in the depletion region due to full ionization, resulting in more uniform and intense electric field lines from the gate potential to ionized acceptors, which improves gate integrity of the channel. Table 4.4 discusses the oxide thickness variation effect on threshold voltage for $L = 0.7$ micron for the different substrate temperatures selected.

Table 4.4 shows that scaling of channel length and oxide thickness increases the threshold voltage standard deviation slightly for the general trend we observe from $T = 100K$ to $T = 500K$. Toward the lower temperature of operations, this increase in standard deviation of threshold voltage is mostly caused by lower oxide thickness range values and the generally higher substrate dopant concentration values. The lower oxide thickness values chosen significantly lower the mean threshold voltage value, and since the scaled MOSFET mean dopant value needs to be higher, the freeze out effect is more for higher mean substrate dopant value. The higher mean threshold voltage value resulting from higher oxide thicknesses in the variation range will contribute lesser

Table 4.4: V_T value for the selected oxide thickness variations from mean value ($L = 0.7\,\mu$m)

Value	tox=36nm	tox=46nm	tox=56nm	tox=66nm	tox=76nm	STDEV
T=100K	0.97254	1.1828	1.3931	1.6034	1.8137	0.332501
T=150K	0.91262	1.1156	1.3187	1.5217	1.7247	0.321012
T=200K	0.84652	1.0412	1.236	1.4307	1.6254	0.307889
T=250K	0.77499	0.96042	1.1458	1.3313	1.5167	0.293191
T=300K	0.69826	0.87336	1.0485	1.2236	1.3987	0.276876
T=350K	0.61622	0.77987	0.9435	1.1072	1.2708	0.258752
T=400K	0.5284	0.6793	0.8302	0.981	1.132	0.238578
T=450K	0.43389	0.57043	0.7069	0.8435	0.98	0.215871
T=500K	0.331	0.45106	0.5711	0.6912	0.8112	0.189822

deviation in the standard deviation computed. We also observe the general trend of standard deviation that it increases to lower extremity of substrate temperature from $T = 300$K and decreases to higher extremity of substrate temperature.

Next we consider the impact of negative substrate bias variation on threshold voltage standard deviation considering the chosen range of substrate temperatures. Negative substrate bias widens the depletion region and surface band bending and increases the mean threshold voltage value. One of the impacts of additional widening of depletion region and associated surface band bending is that even in the midst of sparse ionized dopants in the lower operating substrate temperatures, a gate vertical field can penetrate deeper into the depletion region, and hence variations in threshold voltage will be more uniform. Also the mean value of threshold voltage is enhanced due to substrate bias, and as a result, the standard deviation will be improved and will be less compared to the cases we observed when the threshold voltage became lower owing to reduced oxide thicknesses.

From Table 4.6, we observe that due to scaling of channel length, standard deviation improves and decreases slightly. Overall the trend of Table 4.5 is more or less preserved in Table 4.6, and since surface band bending and channel depletion width can be separately controlled by substrate bias putting negative voltage on it, we can get rid of the dopant variation effect and oxide thickness variation effect on threshold voltage, which are also dependent on surface band bending and depletion width. Inspection of Tables 4.5 and 4.6 also illustrates the fact that negative substrate bias engineering is indeed a more potent and viable option device engineers have consistently employed over the years to tackle the variations of mean threshold voltage drift from die to die and within die MOSFET device fabrication. Next we present Tables 4.7 and 4.8 that highlight the simulation outcomes at different temperatures for positive substrate bias variation on threshold voltage for $L = 1$ micron and $L = 0.7$ micron channel length n-MOSFETs.

Table 4.5: V_T value for the selected negative substrate bias voltages variations ($L = 1\ \mu$m)

Value	Vsb=0.25V	Vsb=0.50V	Vsb=0.75V	Vsb=1.0V	Vsb=1.25V	STDEV
T=100K	1.7349	1.8746	2.0033	2.1235	2.2365	0.198148
T=150K	1.6546	1.7981	1.9299	2.0525	2.1675	0.202612
T=200K	1.566	1.7142	1.8495	1.9749	2.0923	0.207872
T=250K	1.4701	1.6237	1.7632	1.8918	2.0118	0.213947
T=300K	1.3673	1.5273	1.6715	1.8039	1.9269	0.220995
T=350K	1.2574	1.4251	1.5749	1.7114	1.8378	0.229169
T=400K	1.1402	1.3171	1.4733	1.6147	1.7449	0.238721
T=450K	1.0149	1.203	1.3668	1.5137	1.6481	0.249921
T=500K	0.8802	1.0824	1.2551	1.4085	1.5478	0.263401

Table 4.6: V_T value for the selected negative substrate bias voltages variations ($L = 0.7\ \mu$m)

Value	Vsb=0.25V	Vsb=0.50V	Vsb=0.75V	Vsb=1.0V	Vsb=1.25V	STDEV
T=100K	1.5687	1.692	1.8057	1.9119	2.0118	0.175041
T=150K	1.4978	1.6243	1.7406	1.8487	1.9503	0.178743
T=200K	1.4195	1.5498	1.669	1.7796	1.8831	0.183128
T=250K	1.3347	1.4695	1.5922	1.7054	1.8111	0.188169
T=300K	1.2439	1.384	1.5106	1.6269	1.7351	0.193992
T=350K	1.1471	1.2934	1.4246	1.5444	1.6554	0.200727
T=400K	1.0441	1.1979	1.3343	1.4581	1.5722	0.208507
T=450K	0.9344	1.0973	1.2398	1.3681	1.4857	0.21761
T=500K	0.8173	0.9912	1.141	1.2745	1.3961	0.228407

From Table 4.7, it is clear that positive substrate bias of low applied voltage magnitude significantly lowers the variations in the threshold voltage, i.e., standard deviation. This is due to the minority electron injection into the surface channel close to n^+ source due to substrate to n^+ source forward bias impact of inherent diode. Additionally this effect reduces the mean absolute threshold voltage at all temperatures, and therefore, the standard deviation outcome is also lower. Hence, positive substrate bias can be used at present day low voltage MOSFET technology, as the obtained threshold voltage standard deviation is mostly very low over the temperature range from 100K to 400K. At higher temperature, due to the positive substrate bias, threshold voltage goes down very rapidly, and the substrate (p) and source (n^+) forward biased diode will also exhibit more non-ideality, owing to the substrate temperature rise. As a result of both of these

Table 4.7: V_T value for the selected positive substrate bias voltages variations ($L = 1\,\mu$m)

Value	Vsb=-0.30V	Vsb= -0.20V	Vsb= -0.15 V	STDEV
T=100K	1.3699	1.4443	1.4799	0.0561289
T=150K	1.2748	1.353	1.3902	0.0589014
T=200K	1.1677	1.2508	1.2901	0.0624925
T=250K	1.0485	1.138	1.1799	0.0671216
T=300K	0.9159	1.014	1.0594	0.0733451
T=350K	0.7667	0.8771	0.9271	0.0820735
T=400K	0.5941	0.7234	0.7804	0.0954596
T=450K	0.3801	0.5459	0.6141	0.1203446
T=500K	0.0198	0.3252	0.4162	0.2076388

phenomena, threshold voltage standard deviation steeply peaks around $T = 450$K or higher even for very low positive substrate bias voltage ranges.

Table 4.8: V_T value for the selected positive substrate bias voltages variations ($L = 0.7\,\mu$m)

Value	Vsb=-0.30V	Vsb=-0.25V	Vsb=-0.20V	Vsb=-0.15V	Vsb=-0.10V	STDEV
T=100K	1.2128	1.245	1.2763	1.3067	1.3362	0.048785
T=150K	1.1306	1.1644	1.1972	1.2289	1.2596	0.051001
T=200K	1.0381	1.074	1.1086	1.142	1.1743	0.053834
T=250K	0.9356	0.9742	1.0112	1.0467	1.0808	0.057397
T=300K	0.8223	0.8645	0.9045	0.9427	0.9793	0.062037
T=350K	0.696	0.7432	0.7875	0.8293	0.8691	0.068392
T=400K	0.5523	0.6073	0.6578	0.7048	0.7489	0.077661
T=450K	0.3808	0.4499	0.5107	0.5656	0.616	0.092853
T=500K	0.1432	0.2525	0.3352	0.4046	0.4656	0.126888

Table 4.8 shows the threshold voltage values at different temperatures chosen for the range of low positive substrate voltage bias and the associated standard deviation values. Due to channel length scaling, the standard deviation has overall decreased over the selected substrate temperature range. The physical explanation of the sudden steep rise of standard deviation for positive substrate biases when substrate temperature is 450K or higher has been given above when explaining Table 4.7 data. In comparisons of impact of dopant concentration variation, oxide thickness variation, negative substrate bias variation, and positive substrate bias variation, we see that with regard to lowest values of standard deviations of threshold voltage over the temperature range

when scaling of MOSFET channel length is considered negative substrate bias variation and positive substrate bias variation are the most efficient and robust device centric engineering ways to control threshold voltage value and its variation at a specific substrate temperature operation which should focus on the low temperature data. Finally, we show the bar chart for the scaled MOSFET threshold voltage standard deviation for $L = 1$ micron and $L = 0.7$ micron, respectively.

From Figure 4.1, it is clear that the least standard deviation of threshold voltage V_T is caused by positive substrate bias variation over the substrate temperature range. Then second most stan-

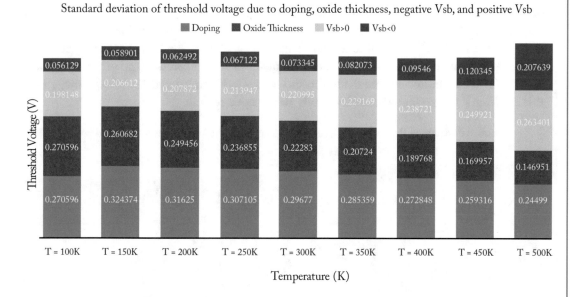

Standard deviation of threshold voltage due to doping, oxide thickness, negative Vsb, and positive Vsb

Figure 4.1: Bar diagram showing the variation in threshold voltage for different temperatures for variations in (i) substrate doping, (ii) oxide thickness, (iii) negative substrate bias, and (iv) positive substrate bias for n-MOSFET of channel length $= 1$ micron.

dard deviation of V_T is caused by variation in oxide thickness. The third most standard deviation in V_T is caused by negative substrate bias which the design engineers commonly choose to set the V_T of different circuits. The most deviation in V_T is caused by substrate doping variation. Therefore, in order to determine threshold voltage variability window, it is the substrate doping variation induced standard deviation of threshold voltage in the range $T = 100K$ to $T = 300K$ which we have to take into account. Based on Figure 4.1, if we average the five sample standard deviation values of V_T extracted from variation in substrate doping concentration for $T = 100K$ to 300K, it is 0.3031 V. Now since the threshold voltage scales by $\sqrt{k}$ where k is the scaling parameter for channel length, we see that from 1 micron gate length or 1,000 nm gate length to the current scaling scenario of 10 nm, $k = 10/1,000 = 0.01$ or $\sqrt{k} = 0.1$. Hence, at channel length of 10 nm, the expected standard deviation derived from largest variation is $= 0.1 \times 0.3031$ V $= 0.0303$ V or 30.3 mV. Through device modeling it has been computed that for 10 nm gate length, possibly fabricated as gate all around nanowire MOSFET, threshold voltage is of the order of 90 mV to 100 mV. We can see the variation tolerance window of V_T is a substantial portion of mean threshold voltage value at the 10 nm node as caused by different process parameter variations. This is indeed a large window that we have to allow to maintain the die V_T consistent with substrate doping variations, although it is true that since at lowered temperature operation, the body is intrinsic because of the added advantage of mobility boost, this window of operation might need to be maintained for other process parameters variation, such as oxide thickness variation or negative substrate bias variations which will reduce the tolerance window of V_T to within 25 mV or equal to 20 mV. This value is around 20–25% of mean V_T for 10 nm node, taking it to be 100 mV. Stringent control of substrate temperature and process parameters variation will be required so that die to die and intra-die V_T (min) and V_T (max) remain within this 20–25 mV range when the device enters 10 nm node regime.

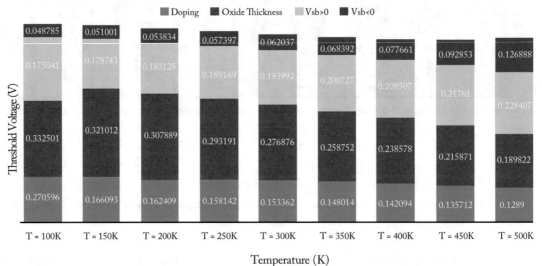

Figure 4.2: Bar diagram showing the variation in threshold voltage for different temperatures for variations in (i) substrate doping, (ii) oxide thickness, (iii) negative substrate bias, and (iv) positive substrate bias for n-MOSFET of channel length = 0.7 micron.

CHAPTER 5

Advantage of Lower Substrate Temperature MOSFET Operation to Minimize Short Channel Effects and Enhance Reliability

5.1 LOW SUBSTRATE TEMPERATURE MOSFET MODELING BENEFITS IN CONSIDERATION OF SHORT CHANNEL EFFECTS

The previous sections with simulation outcomes provided a detailed presentation of long channel MOSFET performance advantages with regard to threshold voltage control utilizing the low substrate temperature operation. Although simulation results yet to be conducted on scaled MOSFET when channel length is less than 0.65 micron when the MOSFET enters into short channel regime, in this section the authors provide a general preview of how a low temperature MOSFET operation can efficiently overcome the barriers to superior device performance enforced by typical short channel effects. One of the potent short channel effects is the threshold voltage roll off caused by increased 2-dimensional charge sharing by source to body and drain to body junctions as channel length is scaled diminishing the distance between source and drain junction contacts. While at substrate temperatures close to 100K incomplete ionization will reduce the active dopants of the substrate, and hence the source-body and drain-body junctions will be wider and enhanced, which gives the indication that threshold voltage roll off effect will be worse, the actual mean threshold voltage indeed is at a higher value compared to $T = 300K$, as we can identify from the $V_T - T$ relationship. Therefore, $\frac{\Delta V_T}{V_T}$ will not increase or more or less remain at the same proportionate value when this measurement is extended from $T = 300K$ to $T = 100K$ as channel length L is reduced. As given by Yuan Taur in his textbook,

$$\Delta V_t\,(SCE) = \frac{qN_A W_{dm}}{C_{ox}} \left(\sqrt{1 + \frac{2W_{dm}}{x_j}} - 1 \right) \frac{x_j}{L}.$$

In this equation, due to lower substrate temperature operation, W_{dm} is larger than room temperature W_{dm}. Also W_{dm} is inversely proportional to $(N_A)^{1/2}$. Considering ultra shallow junction thickness x_j, the expression in the first bracketed term under the square root can be approximated as $\frac{2W_{dm}}{x_j}$ and the equation can be rewritten as,

$$\Delta V_t\,(SCE) = \frac{\sqrt{2}q\,N_A\,W_{dm}^{3/2}}{C_{ox}}\frac{\sqrt{x_j}}{L} - \frac{q\,N_A\,W_{dm}}{C_{ox}}\frac{x_j}{L}.$$

As $W_{dm} \propto N_A^{-0.5}$, $W_{dm}^{3/2} \propto N_A^{-0.75}$ and the first term of the above equation is proportional to $(N_A)^{0.25}$ for a known selected x_j as per the scaling rule for channel length L. The subtractive term of the above equation is proportional to $(N_A)^{0.5}$. Hence, we see that with the incomplete ionization effect, as N_A is more than 20% reduced in the vicinity of $T = 100K$ from its room temperature 100% ionization value, the first term is reduced substantially whereas the subtractive term as a product term first gets decreased by the factor $(N_A)^{0.5}$ due to carrier freeze out, and also gets decreased by $\frac{x_j}{L}$ in the product term as x_j is typically lower than L in the short channel regime. Therefore, at the vicinity of $T = 100K$, the roll off factor ΔV_T will either be reduced less than its room temperature counterpart or remain unaffected by the additional decrease of the subtractive term boosting the overall ΔV_t when channel length is not so aggressively scaled. When channel length is much more aggressively scaled, x_j/L value is close to unity or slightly less than unity and then the subtractive term of the above equation when operated at $T = 100K$ will be dependent on the $(N_A)^{0.5}x_j/L$ product term which will not decrease as sharply as was the case when channel length L is larger albeit still in the short channel regime. The increase in the subtractive term will then decrease the overall $\Delta V_T(SCE)$ when channel length is aggressively scaled and the substrate temperature is in the vicinity of $T = 100K$ from its room temperature operation. The above analysis thus shows that overall the threshold voltage roll off effect will be reduced or unaffected by lowering the substrate temperature operation of scaled n-MOSFET.

As the channel width is also reduced to constrict the MOSFET layout area for a high density integrated circuit, we know that narrow width increases the $\Delta V_T(SCE)$ or the mean V_T. Hence, with a reduced substrate temperature operation, the already enhanced V_T due to lower T will further be aggravated by the increase due to narrow width of the gate area. The only improvement comes from the total depletion charge as computed by depletion charge/gate area multiplied by total gate area. In a 2-D arrangement this leaves behind additional total charge near the source unaccounted for by gate field. But this charge is proportional to $(N_A)^{0.5}$ and this value decreases at reduced temperature due to career freeze out. Hence, the total charge, unaccounted by the gate active area, which is responsible for V_T increase, will generate less up shift of V_T as its total value is reduced at lower substrate temperature operation. Still narrow gate width will have a detrimental effect on threshold voltage at lower operating temperature, as this effect always will enhance V_T, which is not amenable to low voltage scaling that the designers must meet at every scaled node operation of n-MOSFETs.

A very important short channel effect that exist, in a naturally scaled n-MOSFET device is the drain-induced-barrier-lowering or (DIBL) that influences the drain current in the weak inversion region bias of gate voltage and drain potential and also affects drain output resistance and device transconductance. Due to DIBL, the drain side electric field due to drain potential and drain-to-source-side depletion region extends toward the source due to shortened source to drain junction distance, and as a result of this, the gate to source side barrier for electrons to be injected into the channel from source contact is reduced. This means at lower operating gate voltage, there is sufficient inversion of the channel, hence the drain current as a function of gate voltage in the subthreshold and weak inversion regime shifts toward the left as a result of positive DIBL. Due to the increased threshold voltage at lower operating substrate temperature of n-MOSFET, the natural voltage based scaling has to be relaxed to a proportionate amount for aggressively scaled MOSFET if the device is operated near $T = 100K$. As a result of this the scaled voltage requirement on gate voltage and drain voltage will be proportionally higher than what the conventional scaling rule will predict. Since we have earlier stressed that in order to take advantage of peak mobility at lowered substrate temperature operation of the n-MOSFET, the channel doping has to be near intrinsic value. The drain side drain to substrate junction width will be more wide than when compared with $T = 300K$ junction width. As a result the proportionally scaled higher drain potential as derived from lower substrate temperature operation of n-MOSFET will be sufficient to affect the gate-to-source side barrier potential much more strongly through drain to source electric field penetration when lower than $T = 300K$ substrate temperature operation is considered. Hence, DIBL will be naturally worse for substrate temperatures less than 300K and near 100K. There are some important probing considerations though as we make this assessment for DIBL at lower operating temperature. First the bulk potential formed at the reverse biased drain to substrate junction is quite high due to orders of magnitude lower intrinsic carrier concentration of silicon at lower substrate temperature. Hence, the barrier potential at drain-substrate junction will be considerably higher at lower operating substrate temperature than the same for room temperature. With the bias potential of drain contact considered, the barrier voltage of the reverse biased drain-to-substrate junction will be further extended. This barrier potential should prevent some of the electric field penetration from drain to the source contact underneath the gate through the broad depletion region and this will hinder the cumulative impact of drain field lines extending toward the source and interacting with gate to source barrier as built up by wider depletion width of drain to substrate junction as explained.

Next we are going to discuss the impact of lower operating temperature operation on subthreshold drain leakage current and the gate leakage current when the MOSFET is off or in an idle state. The subthreshold drain current has the following expression:

$$I_{ds} = \mu_{eff} \frac{W}{L} \sqrt{\frac{\varepsilon_s q N_A}{2\psi_s}} \left(\frac{kT}{q}\right)^2 \left(\frac{n_i}{N_A}\right)^2 e^{\frac{q\psi_s}{kT}} \left(1 - e^{\frac{-qV_{ds}}{kT}}\right).$$

From deep scrutiny of the above equation we come across the following understanding of the lower substrate temperature impact on subthreshold leakage current. First the mobility μ_{eff} will be higher at a lower operating substrate temperature set by mobility peak by phonon scattering and slight decrease in ionized impurity scattering due to freeze out of acceptor ions and also increased mean free path of electrons before a scattering can take effect at lower operating temperature. Hence, this mobility increase will obviously serve as a precursor for subthreshold leakage current increase at lower temperature. The term in the square root of the above equation will decrease to slightly as near subthreshold operation at lower T; the surface band bending ψ_s is higher and N_A is lower as explained in preceding subsections. Then this term will certainly neutralize some of the ascending impact on subthreshold leakage current as caused by μ_{eff}. The first squared bracketed term is the thermal voltage squared which will be reduced reciprocally squared as T is reduced. The second square bracketed term as used in the above equation will be reduced substantially as temperature is chosen near $T = 100K$ due to orders of magnitude lower intrinsic carrier concentration which reduces leakage current value in several negative exponents of ten. The exponential term right next to this term will increase the drain current subthreshold leakage current to some higher aggregate number, but its capability to engender several positive powers of ten on the final value will be still lesser than the negative exponents of ten that the square bracketed term containing ratio of intrinsic carrier concentration to substrate doping will generate. The final product component term of the above equation is critical for scaled drain potential's impact on subthreshold leakage current. Naturally at $T = 300K$, with voltage scaling, V_{ds} is reduced to the extent that its impact on drain current coming out of the final product term component will be a fraction of unity and will naturally reduce the final subthreshold leakage current value but at reduced $T = 100K$, the requirement on drain voltage due to V_T increase is higher than the scaled voltage will infer. As a result the last term in the above equation will be more close to unity and hence will not reduce the drain subthreshold leakage current any further when T is near 100K. After analyzing the overall impact of the terms in the product of the above equation, we can safely say that the subthreshold leakage current will be reduced at lower operating temperature close to 100K due to orders of magnitude negative exponent of ten contributions from $\left(\frac{n_i}{N_A}\right)^2$ term.

The gate current when the MOSFET is switched off is for three components: (i) Fowler Nordheim tunneling current due to triangular oxide energy barrier, (ii) direct tunneling current due to trapezoidal energy barrier, and (iii) trap-assisted-tunneling (TAT) due to Coulomb field induced electron hopping from one trap site to other trap sites. The first two current components are exponentially dependent on (i) oxide peak energy to semiconductor conduction energy band barrier potential ϕ_b (eV) which at $T = 300K$ or room temperature is around 3.2 eV and (ii) electron effective or tunneling mass m^* which decreases with increasing temperature from $T = 300K$ and increases from reduced substrate temperature from $T = 300K$ operation. As the substrate temperature T is reduced from room temperature, the band gap of silicon E_g widens and due to much deeper depletion region, few electrons are available at the conduction band minimum at the surface to surmount the oxide barrier and tunnel to gate potential. Although taking the valence band electron energy as reference E_V (eV), widening of E_G from its near 1.12 eV value at $T = 300K$ will decrease the ϕ_b (eV) and consequently electron tunneling barrier from conduction band over the oxide field to the gate electrode potential is reduced. But since at $T = 100K$ and very small gate voltage the concentration of inversion carrier electron is substantially minimized, the net gate current at $T = 100K$ will be substantially less than the gate current possible at $T = 300K$ at the reduced gate and drain voltage when the MOSFET is near off condition or remains in its idle state. The other component tunneling effective mass also beneficially assists the reduction of gate tunneling current. The effective mass component inversely affects the exponential term contribution to gate current. At reduced $T = 100K$, tunneling mass is enhanced from room temperature and hence the exponential term contribution to gate current will be further reduced in addition to barrier voltage contribution. With more voltage scaling and oxide thickness field scaling, Fowler Nordheim tunneling gradually decomposes to direct tunneling current, and since the direct tunneling current is mostly determined by tunneling barrier width rather than oxide field, this current component will be mostly affected at lower operating temperature by the probability of occupancy of conduction band energy states by inversion electrons responsible for tunneling. Since at lower operating temperature, the number of electron density near the surface of the oxide-substrate is significantly less, the net gate tunneling leakage current due to this current component will be also reduced. The trap-assisted-tunneling (TAT) is governed by the repulsive Coulomb field that results in the vicinity of a trap when the trap is occupied by an electron and the trap is an acceptor. If the Coulomb field is not deep, the electrons through their thermal energy can hop between trap sites and form a conduction path through oxide to the gate terminal. If the depletion region is wider due to the $T = 100K$ operation, there is less probability that an inversion layer electron will be captured by an oxide trap very close to the channel due to significantly lower concentration of electrons in the channel at T near 100K. Furthermore, since less and less number of traps will be occupied by electrons, the trap Coulomb repulsive potential will be much deeper at $T = 100K$, and due to reduced thermal energy induced velocity, the trapped electron will not be able to surmount this Coulomb repulsive field to hop from one trap site to the next. Therefore, at $T = 100K$, the trap-assisted-tunneling (TAT) current will be much less than the same happening at room temperature. Overall the gate leakage current will be less when the substrate operating temperature is reduced to $T = 100K$.

The last current component that we are going to focus on now is source to drain direct punch-through current. Near the limits of scaled channel length, the source and drain contact are in closest proximity than ever and this will trigger the source to drain direct punchthrough current as one of the potential roadblock for reliable MOSFET operation at Moore's Law based scaling evolution. At lower temperature, it is true that due to intrinsic substrate doping, both the source to substrate and drain to substrate junctions are wider than their corresponding at room temperature. As a result there must be design and provision so that at the scaled node of channel length, these two depletion regions do not touch, merge, or overlap. Although the depletion region is wider at $T = 100K$ near source to substrate, the barrier potential of the reverse biased junction instead of being lower turns out to be higher due to much reduced intrinsic carrier concentration and hence the overall reverse biased source to substrate depletion region barrier potential is increased. Hence, at an operating gate voltage, electrons responsible for source to drain punch through current will be injected into the channel in lower concentration compared to $T = 300K$. Therefore, if through dopant engineering, temperature control, drain and gate potential selection, we can control the depletion region widths near the source and drain side such that two depletion regions do not merge, then the source to drain direct punch through current will be lesser for $T = 100K$ compared to $T = 300K$ for the above explanations taken into account.

CHAPTER 6

A Prospective Outlook on Implementation Methodology of Regulating Substrate Temperatures on Silicon Die

6.1   A SHORT OUTLOOK ON IMPLEMENTATION OF LOW SUBSTRATE TEMPERATURE MOSFET MODELING AND CONTROL

Now that we have described the growing impetus behind lower substrate temperature operation of MOSFET for gaining in a host of reliable MOSFET performance metrics including short channel effects for high density integrated circuits, in this subsection we present a practical outlook on how the implementation of substrate temperature control can be ingeniously devised by the IC professionals working industry. In typical versatile logic circuits such as microprocessors, different parts of core blocks as members of a microprocessor logic unit perform different functions run-time based on clock inputs. The logic designers perform typical scheduling, pipelining and parallelism to obtain the required drain drive current requirements, threshold voltage requirements and drain leakage current requirements for all the ICs as part of different core logic blocks. After knowing the required switching related and conduction related maximum drain current drivability and maximum leakage current when a pool of IC blocks need to pass through, we will need to estimate several performance parameters, such as electron mobility, MOSFET threshold voltage, MOSFET drift velocity, etc. For a typical clock synchronization scheduling of different parts of IC blocks, all these device centric parameters require adjustments so that drive current, subthreshold leakage current and gate current requirements can be accomplished within a tolerance from their circuit operating values. On-chip co-fabricated solid state micro-refrigerators first require being integrated to the different core blocks as fabricated in different parts of the silicon die. These micro-refrigerators as residing with the core circuit architecture at respective die location will control the temperature particularly at the lower operating values at these respective die areas and hence control the device parameters adjustments as per detailed modeling highlighted in this E-book. Based on the different device parameter's adjustments as emanated from run-time operation of typical core IC blocks to give rise to suitable maximum drain current variability and maximum leakage current variability, these co-fabricated solid state micro-refrigerators can set the die substrate temperature native to these core IC blocks to the range as to extract the most efficient performance at the lowered substrate temperature operating conditions. A microcontroller based central processing unit will have to pass on the required electrical signals or amplitude voltages to all the co-fabricated micro-refrigerators native to different IC blocks to convert the electrical energy to the specified thermal energy and hence the required temperature that becomes an identity for respective parts of the different die-sections where different arithmetic and logic core units are fabricated as components of the microprocessor circuit. Therefore, the centrally processed microcontroller unit has to be programmed with hardware description language (HDL) such that when synchronized with clock pulses, along with different core assembly IC blocks of different parts of the silicon die, it will generate the required electrical signal amplitudes being channeled to the respective solid state micro-refrigerator micro units to set the die temperature native to IC blocks fabricated in sectional arrangements in a VLSI circuit lay-out. The construction of the micro-refrigerator and its selection of refrigerant or coolant is well researched and a number of excellent articles are available on the miniature solid state micro-refrigerator design from micro-electro-mechanical-systems (MEMS) aspect and taking the dimensions in mm regime. If the variations of required die temperatures at different core IC locations can be minimized and controlled within a narrower range focusing on the improved microprocessor performance with the ever increasing clock speed, the amount of volume overhead arising from the need of co-fabrication of micro-refrigerators will be substantially reduced along with the stringent scheduling requirement as per time-line of clock cycles serving the execution of vast hardware description languages being installed in the centrally processing unit when the required die temperatures are not within narrow ranges from optimal lower substrate temperature chosen.

CHAPTER 7

Summary of Research Results

7.1   SUMMARY OF RESEARCH OUTCOMES

As outlined in the introduction section of this present E-book, the authors have posited lower substrate/lattice temperature based device architectural innovations of what we know today to enable device engineers to keep up with the scaling based evolution particularly at 10 nm and beyond. In order to depict the modeling scenario where a number of device parametric features central to MOSFET threshold voltage, off-state leakage, and drive (on) current computations undergo significant variations at lower substrate temperature operations of a n-MOSFET device, the authors have taken recourse to standard well-cited and reputed textbook modeling equations, particularly the parametric adaptations essential for lower substrate and lattice temperature operation. The authors have systematically performed device physics based analysis of incomplete ionization, a feature of dopant ionization percentage at lower substrate temperature, on different substrate doping concentrations. Particularly, the incomplete ionization effect is severe on very high doping concentrations close to degenerate limit which is required for extremely short channel MOSFET for adequate source to drain junction separation to prevent punch-through. In this summary we also point out that instead of operating at room temperature, if the substrate temperature is within 100K below room temperature, the device substrate doping can be tailored to remain close to intrinsic doping or near $10^{14}/cm^3$ or $10^{15}/cm^3$ which will be less impacted by the effect of incomplete ionization and the details are discussed in the relevant section. One of the parameters that is often overlooked in determination of threshold voltage-substrate temperature profile is the flat-band voltage ϕ_{ms} and its variation with temperature for a reference doping at a reference channel length. This E-book has detailed the proper modeling of ϕ_{ms} with substrate temperature with the profile plot illustrated in the relevant section. The threshold voltage variation with substrate temperature has been very meticulously developed by industry standard analytical modeling equations referenced from the best available textbook in this field. The threshold voltage as a function of substrate temperature has been extracted for a set of substrate doping values in the vicinity of reference substrate doping, gate oxide thickness values in the vicinity of reference oxide thickness, a set of negative substrate biases and for a set of small positive substrate bias. These plots of threshold voltage as a function of substrate temperatures from 100K to 500K will serve as a guiding tool for device engineers to project their device centric optimum threshold voltage at a reference doping based on a reference gate length n-MOSFET for a set of reduced substrate temperature operations. Although the threshold voltage variation with regard to substrate doping and oxide thickness were introduced to very few modeling works in the past, which themselves are not exhaustive, in this summary we would like to point out that the adaptive body bias features that are successfully employed in IC design and verification to adjust die threshold voltage, are now fully equipped through this E-book with threshold voltage-substrate temperature profiles for a set of negative and positive substrate or body biases. The proper and accurate modeling sequences as executed in computations of various parameters contributing to final threshold voltage also reveal important deviations as encountered at temperatures close to 100K and 500K when a straight line decline of threshold voltage as a function of substrate temperature is preserved at most other temperatures in the vicinity of room temperature 300K. The tabular datasets are also provided when discussing threshold voltage variation with substrate temperature. Bar diagrams are inserted to draw the attention of the readers on the distinctive contributions to threshold voltage variation at five different temperatures from 100K to 500K for the contributions emanating from (i) dopant concentration variation, (ii) oxide thickness variation, (iii) negative substrate biases, and (iv) positive substrate biases. Compared to negative substrate bias, positive substrate biases introduce more anomalies in threshold voltage deviations at temperatures close to 100K and 500K. The device physical analyses behind these attributes are systematically documented in the relevant section. From the standard deviation of threshold voltage variation analyzed from the bar diagram chart for reference gate length of 1 micron, a scaled projection of maximum threshold voltage variation tolerance near 10 nm or the current node of industry projection is outlined. From short channel MOSFET second order effects such as threshold voltage roll off, drain-induced-barrier lowering, subthreshold drain leakage current, off-state gate current and source to drain direct punch through current are properly analyzed from the context of using lower substrate temperature operation of scaled MOSFET. The device physical analyses of these short channel effects that impact the reliable performance of the device for prolonged long time operation clearly bolster the added benefits of operating the present day scaled device architecture at reduced substrate temperature within 100K of room temperature 300K. A tentative solid state micro-cooler or refrigerator based IC integration scheme is outlined with their ingenious co-fabrication in different subsections of a silicon die equipped with integrated circuits of different functional forms of a core microprocessor unit. This book will be highly beneficial to device, process, and manufacturing engineers and professionals in trying to optimize the temperature of the substrate lower than room temperature when simultaneously they are designing the proper device architecture to remain on par with the ITRS roadmap for each technology node with the required performance deliverance datasets. Some of the issues that need to be investigated further through experimental and numerical device modeling are the role of reduced substrate temperature operation on high-k/metal gate stack scaled MOSFET performance. We therefore recommend [64–69] for information on high-k based device performance issues. Another important effect that becomes prominent at lower device operating temperature is the growing source, gate, and drain side contact resistances and other resistances present at many interfacial boundaries of scaled MOSFET during device fabrication. It is due to low carrier activation at ohmic contacts at lower substrate temperature leading to contact resistance rise. One possible remedy for this effect can be implementation of a cyclical laser thermal heating or rapid thermal activation of source, drain, and gate metal contact regions during the time the device operating temperature is below 300K and a time-synchronized termination of the laser heating would ensure proper dopant activation enhanced contact resistance and parasitic effects minimization minimally affecting the intrinsic speed of the device.

CHAPTER 8

Conclusion

This book was written for technology students, academicians, and industrialists dealing with the nagging issues of device scaling-based performance maintenance and in certain cases improvement of today's scaled ICs. Although different device architectures are already configured at room substrate/lattice temperature to acquire as much improvements as possible in device metrics such as subthreshold slope, mobility, and drive current, the gamut of their achievability diminishes with scaling entering 10 nm node and below. One of our aims in this book was to employ the substrate temperature feature as an intrinsic control and devising scheme to achieve a range of device parameters metrics improvements that are not possible through device architectural innovation. The authors outlined the modeling equations for the critical evaluating factor, i.e., threshold voltage as a function of temperature for different device and process parameters that engineers use in practice to set threshold voltage of a core IC block and generate multiple threshold voltage schemes. Through graphical illustrations the authors shed light on device physical attributes that underlay these different simulation outcomes of threshold voltage variation with temperature and how knowledge of this profile can be used to control threshold voltage variation within different IC blocks on the same die or on batch manufactured die-to-die variation. In later subsections the authors discussed the emergence of short channel features that can be controllably reduced at lower substrate temperatures. The authors explained how these different die temperatures can be implemented on different core IC sections of a logic microprocessor unit through multi-scheduling spanning executable clock cycles. This book will serve as a useful tool for practicing device engineering professionals and pupils to find new enterprising device architectural solutions blended with masterfully engineered substrate temperature control essentially in the lowered substrate operating regime as we would like to be at pace with the Moore's Law- based shrinking domain.

References

[1] Pierret, R. F., *Semiconductor Device Fundamentals*, Reading, MA: Addison-Wesley, 1996, pages 23–74 and 563–690. 5, 17, 18, 19

[2] Pierret, R. F., *Advanced Semiconductor Fundamentals, Modular Series on Solid state Devices*, Pearson Education Inc., Upper Saddle River, NJ, 2003, pages 87–133 and 175–215. 19

[3] Streetman, B. G. and Banerjee, S. K., *Solid State Electronic Devices*, 6th ed., Upper Saddle River, NJ, Pearson Prentice Hall, 2006.

[4] Taur, Y. and Ning, Tak H., *Fundamentals of Modern VLSI Devices*, Cambridge University Press, 1998, pages 112–221. 6

[5] Sze, S. M., *VLSI Technology*, McGraw Hill, NY, 1983. 17

[6] Sze, S. M. and Ng, Kwok K., *Physics of Semiconductor Devices*, John Wiley & Sons Inc., NJ, 2007.

[7] Sze, S. M., *Semiconductor Devices: Physics and Technology*, John Wiley & Sons Inc., NJ, 1985. 17

[8] Shur, M., *Physics of Semiconductor Devices*, Prentice and Hall, 1st ed., 1990. DOI: 10.1063/1.2810727.

[9] Shur, M., *Introduction to Electronic Devices*, John Wiley & Sons Inc, NJ, 1996.

[10] Neamen, Donald A., *Semiconductor Physics and Devices: Basic Principles*, 4th ed., McGraw Hill Inc, NY, pages 106–191 and 371–490. 24

[11] Tsividis, Yannis P., *Operation and Modeling of the MOS Transistor*, McGraw Hill International Editions, 1988, pages 35–216.

[12] E-book on Principles of Semiconductor Devices, Bart Zeghbroeck.

[13] Colinge, J. P. and Colinge, C. A., Physics of Semiconductor Devices, Springer Science+Business Media Inc., 2002.

[14] Schroder, D. K., *Advanced MOS Devices, Modular Series on Solid State Devices*, Vol. VII, Addison-Wesley, Reading, MA, 1987. 5

[15] Barber, H. D., *Effective Mass and Intrinsic Concentration in Silicon, Solid State Electronics*, Pergamon Press, 1967, Vol. 10, pages 1039–1051. DOI: 10.1016/0038-1101(67)90122-0. 18

[16] Vankemmel, R., Scchoenmaker, W., and Meyer, K. De., A Unified Wide Temperature Range Model for the Energy Gap, The Effective Carrier Mass and Intrinsic Concentration in Silicon, *Solid State Electronics*, Vol. 36, No. 10, 1993, pages 1379–1384. DOI: 10.1016/0038-1101(93)90046-s. 18

[17] Varshni, Y. P., Temperature Dependence of the Energy Gap in Semiconductors, *Physica*, Vol. 34, Issue 1, 1967, pages 149–154. DOI: 10.1016/0031-8914(67)90062-6. 18

[18] O'Donnell, K. P. and Chen, X., Temperature Dependence of Semiconductor Band Gaps, *Applied Physics Letters*, Vol. 58, No. 25, 1991, pages 2924–2926. DOI: 10.1063/1.104723. 18

[19] Taur, Y., Buchanan, Douglas A., Chen, Wei, Frank, David J., Ismail, Khalid E., Lo, Shih-Hsien, Sai-Halasz, George, A., Viswanathan, Raman G., Wann, Hsing-Jen C., Weind, Shalom J., and Wong, Hon-Sum, CMOS Scaling into the Nanometer Regime, *Proceedings of the IEEE*, Vol. 85, No. 4, 1997, pages 486–504. DOI: 10.1109/5.573737. 6

[20] Frank, David J., Dennard, Robert J., Nowak, E., Solomon, Paul M., Taur, Y., and Wong H-S., Device Scaling Limits of Si MOSFETs and Their Application Dependencies, *Proceedings of the IEEE*, Vol. 89, No. 3, 2003, pages 259–288. DOI: 10.1109/5.915374. 7

[21] Frank, David J. and Taur, Y., Design Considerations for CMOS near the Limits of Scaling, *Solid State Electronics*, Vol. 46, 2002, pages 315–320. DOI: 10.1016/s0038-1101(01)00102-2.

[22] Wong, Hon-Sum P., Frank, David J., Solomon, Paul M., Wann, Clement H. J., and Welser, Jeffrey J., Nanoscale CMOS, *Proceedings of the IEEE*, Vol. 87, No. 4, 1999, pages 537–570. DOI: 10.1109/5.752515. 7

[23] Wong, Hon-Sum P., Beyond The Conventional Transistor, *IBM Journal on Research and Development*, Vol. 46, No. 2/3, 2002, pages 133–167. DOI: 10.1147/rd.462.0133. 7

[24] Taur, Y., CMOS Design Near The Limit of Scaling, *IBM Journal On Research and Development*, Vol. 46, No. 2/3, 2002, pages 213–222. DOI: 10.1147/rd.462.0213. 11

[25] Dennard, Robert H., Gaensslen, Fritz H., Yu, Hwa-Nien, Rideout, V. Leo, Bassous, Ernest, and Leblanc, Andre R., Design of Ion-Implanted MOSFETs with Very Small Physical Dimensions, *IEEE Journal of Solid State Circuits*, Vol. SC-9, No. 5, 1974, pages 256–268. DOI: 10.1109/jssc.1974.1050511. 9

[26] Baccarani, G., Wordeman, M. R., and Dennard, R. H., Generalized Scaling Theory and Its Application to a 1/4 micrometer MOSFET Design, *IEEE Transactions on Electron Devices*, Vol. 31, Issue 4, 1984, pages 452–462. DOI: 10.1109/t-ed.1984.21550.

[27] Haensch W., Nowak E. J., Dennard R. H., Solomon P. M., Bryant A., Dokumaci O. H., Kumar A., Wang X., Johnson J. B. and Fischetti M. V., Silicon CMOS Devices Beyond Scaling, *IBM Journal of Research and Development*, Vol. 50, No. 4/5, 2006, pages 339–361. DOI: 10.1147/rd.504.0339.

[28] Davari, B., Dennard, Robert H., and Shahidi, Ghavam G., CMOS Scaling for High Performance and Low Power-The Next Ten Years, *Proceedings of the IEEE*, Vol. 83, No. 4, 1995, pages 595–606. DOI: 10.1109/5.371968. 9

[29] Sun, J-Y C., Taur, Y., Dennard, Robert H., and Klepner, S. P., Submicrometer Channel CMOS For Low Temperature Operation, *IEEE Transactions on Electron Devices*, Vol. 34, Issue 1, 1987, pages 19–27. DOI: 10.1109/t-ed.1987.22881. 7

[30] Sai-Halasz, G. A., Wordeman, M. R., Kern, D. P., Ganin, E., Rishton, S., Zicherman, D. S., Schmid, H., Polcari, M. R., Ng, H. Y., Restle, P. J., Chang, T. H. P., and Dennard, R. H., Design and Experimental Technology for 0.1 μm Gate-Length Low Temperature Operation FET's, *IEEE Electron Device Letters*, Vol. 8, Issue 10, 1987, pages 463–466. DOI: 10.1109/edl.1987.26695. 7

[31] Dennard, Robert H., Cai, J., and Kumar, A., A Perspective on Today's Scaling Challenges and Possible Future Directions, *Solid State Electronics*, Vol. 51, Issue 4, 2007, pages 518–525. DOI: 10.1016/j.sse.2007.02.004.

[32] Dennard, R. H., Gaensslen, F. H., Walker, E. J., and Cook, P. W., 1 μm MOSFET VLSI Technology: Part II-Device Designs and Characteristics for High-Performance Logic Applications, *IEEE Transactions on Electron Devices*, Vol. 26, Issue 4, 1979, pages 325–333. DOI: 10.1109/t-ed.1979.19431.

[33] Dhar, S., Pattanaik, M., and Rajaram, P., Advancement in Nanoscale CMOS Device Design En Route to Ultra Low Power Applications, *VLSI Design*, Hindwai Publishing Corporation, Volume 2011, pages 1–19. DOI: 10.1155/2011/178516. 7, 9

[34] Kim, Y., Review Paper: Challenges for Nanoscale MOSFETs and Emerging Nanoelectronics, *Trans. Electr. Electron. Mater.*, 10(1), 21, 2009, pages 23–41. DOI: 10.4313/teem.2010.11.3.093.

[35] Roy, K., Mukhopadhyay, S., Mahmoodi, and Meimand H., Leakage Current Mechanisms and Leakage Reduction Techniques in Deep Submicrometer CMOS Circuits, *Proceedings of the IEEE*, Vol. 91, No. 2, 2003, pages 305–327. DOI: 10.1109/jproc.2002.808156. 8

[36] Iwai, Hiroshi, Roadmap For 22 nm and Beyond, *Microelectronic Engineering*, Vol. 86, Issues 7–9, 2009, pages 1520-1528.

[37] Skotnicki, T., Hutchby, James, A., King, Tsu-Jae, Wong, H.-S. P., and Boeuf, F., The End of CMOS Scaling: Toward The Introduction of New materials and Structural Changes to Improve MOSFET Performance, *IEEE Circuits & Devices Magazine*, 2005, pages 16–26. DOI: 10.1109/MCD.2005.1388765.

[38] Narendra, Siva G., Challenges and Design Choices in Nanoscale CMOS, *ACM Journal on Emerging Technologies in Computing Systems*, Vol. 1, No. 1, 2005, pages 7–49. DOI: 10.1145/1063803.1063805. 6

[39] Lin, S. and Banerjee K., Cool Chips: Opportunities and Implications for Power and Thermal Management, *IEEE Transactions on Electron Devices*, Vol. 55, No. 1, 2008, pages 245–255. DOI: 10.1109/ted.2007.911763. 14

[40] Shakouri, A. and Zhang Y., On-Chip Solid State Cooling for Integrated Circuits Using Thin-Film Microrefrigerators, *IEEE Transactions on Components and Packaging Technologies*, 28, 2005, pages 65–69. DOI: 10.1109/tcapt.2005.843219. 7, 15

[41] Gaensslen, Fritz H., Rideout, V. Leo, Walker, E. J., and Walker, John J., Very Small MOSFETs For Low Temperature Operation, *IEEE Transactions on Electron Devices*, Vol. 24, No. 3, 1977, pages 218–229. DOI: 10.1109/t-ed.1977.18712.

[42] Semenov, O., Vassighi, A., and Sachdev, M., Impact of Technology Scaling on Thermal Behavior of Leakage Current in Sub-Quarter Micron MOSFETs: Perspective of Low Temperature Current Testing, *Microelectronics Journal*, Vol. 33, 2002, pages 985–994. DOI: 10.1016/s0026-2692(02)00071-x. 7

[43] Chang, L. and Hu, C., MOSFET Scaling into The 10 nm Regime, *Superlattices and Microstructures*, Vol. 28, No. 5/6, 2000, pages 351–355. DOI: 10.1006/spmi.2000.0933. 11

[44] Kim, Nam S., Austin, T., Blaauw, D., Mudge, T., Flautner, K., Hu, Jie S., Irwin, Mary J., Kandemir, M., and Narayanan V., Leakage Current: Moore's Law Meets Static Power, Computer, *IEEE Computers Society*, 2003, pages 68–75. DOI: 10.1109/mc.2003.1250885. 8

[45] Natori, K., Compact Modeling of Ballistic Nanowire MOSFETs, *IEEE Transactions on Electron Devices*, Vol. 55, No. 11, 2008, pages 2877–2885. DOI: 10.1109/ted.2008.2008009.

[46] Chang, L., Frank, David J., Montoye, Robert K., Koester, Steven J., Ji, Brian L., Cotens, Paul W., Dennard, Robert H., and Haensch W., Practical Strategies for Power-Efficient Computing Technologies, *Proceedings of the IEEE*, Vol. 98, No. 2, 2010, pages 215–236. DOI: 10.1109/jproc.2009.2035451.

[47] Keyes, Robert W., Fundamental Limits of Silicon Technology, *Proceedings of the IEEE*, Vol. 89, No. 3, 2001, pages 227–239. DOI: 10.1109/5.915372.

[48] Plummer, James D. and Griffin, Peter B., Materials and Process Limits in Silicon VLSI Technology, *Proceedings of the IEEE*, Vol. 89, No. 3, 2001, pages 240–257. DOI: 10.1109/5.915373.

[49] Wann, Clement H., Noda, K., Tanaka, T., Yoshida, M., and Hu C., A Comparative Study of Advanced MOSFET Concepts, *IEEE Transactions on Electron Devices*, Vol. 43, No. 10, 1996, pages 1742–1753. DOI: 10.1109/16.536820.

[50] Ohmi, T., Sugawa, S., Kotari, K., Hirayma, M., and Morimoto, A., New Paradigm of Silicon Technology, *Proceedings of the IEEE*, Vol. 89, No. 3, 2001, pages 394–412. DOI: 10.1109/5.915381.

[51] Markovic, D., Wang, Cheng C., Alarcon, Louis P., Liu, Tsung-Te, and Rabaey, Jan M., Ultra Low Power Design in Near Threshold Region, *Proceedings of the IEEE*, Vol. 98, No. 2, 2010, pages 237–252. DOI: 10.1109/jproc.2009.2035453.

[52] Dreslinski, Ronald G., Wieckowski, M., Blaauw, D., Sylvester, D., and Mudge T., Near-Threshold Computing: Reclaiming Moore's Law Through Energy Efficient Integrated Circuits, *Proceedings of the IEEE*, Vol. 98, No. 2, 2010, pages 253–266. DOI: 10.1109/jproc.2009.2034764.

[53] Gupta, Sumeet K., Raychowdhury, A., and Roy, K., Digital Computation in Subthreshold Region for Ultra-Low Power Operation: A Device-Circuit-Architecture Codesign Perspective, *Proceedings of the IEEE*, Vol. 98, No. 2, 2010, pages 160–190. DOI: 10.1109/jproc.2009.2035060.

[54] Vitale, Steven A., Wyatt, Peter W., Checka, N., Kedzierski, J., and Keast, Craig L., FD-SOI Process Technology for Subthreshold Operation Ultra-Low Power Electronics, *Proceedings of the IEEE*, Vol. 98, No. 2, 2010, pages 333–342. DOI: 10.1149/1.3570794. 7

[55] Ronen, R., Mendelson, A., Lai, K., Lu, Shih-Lieu, Pollack, F., and Shen, John P., Coming Challenges in Microarchitecture and Architecture, *Proceedings of the IEEE*, Vol. 89, No. 3, 2001, pages 325–340. DOI: 10.1109/5.915377.

[56] Bryant, Randal E., Cheng, Kwang-Ting, Kahng, Andrew B., Kentzer, K., Maly, W., Newton, R., Pileggi, L., Rabaey, Jan M., and Vincentelli, Alberto S., Limitations and Challenges of Computer-Aided Design Technology for CMOS VLSI, *Proceedings of the IEEE*, Vol. 89, No. 3, 2001, pages 341–365. DOI: 10.1109/5.915378.

[57] Doering, R. and Nishi, Y., Limits of Integrated Circuit Manufacturing, *Proceedings of the IEEE*, Vol. 89, No. 3, 2001, pages 375–393. DOI: 10.1109/5.915380.

[58] Davis, Jeffrey A., Venketasan, R., Kaloyeros, A., Beylansky, M., Souri, Shukri J., Banerjee, K., Saraswat, Krishna C., Rahman, A., Reif, R. and Meindl, James D., Interconnect Limits on Gigascale Integration (GSI) in the 21st Century, *Proceedings of the IEEE*, Vol. 89, No. 3, 2001, pages 305–324. DOI: 10.1109/5.915376. 11

[59] Yeo, Yee-Chia, King, Tsu-Jae, and Hu C., MOSFET Gate Leakage Current Modeling and Selection Guide for Alternative gate Dielectrics Based on Leakage Considerations, *IEEE Transactions on Electron Devices*, Vol. 50, No. 4, 2003, pages 1027–1035. DOI: 10.1109/ted.2003.812504. 10

[60] Lu, W., Xie, P., and Lieber, Charles M., Nanowire Transistor Performance Limits and Applications, *IEEE Transactions on Electron Devices*, Vol. 55, No. 11, 2008, pages 2859–2876. DOI: 10.1109/ted.2008.2005158. 11

[61] Campera, A., Iannaccone, G., and Felice, C., Modeling of Tunnelling Currents in Hf-based Gate Stacks as a Function of Temperature and Extraction of Material Parameters, *IEEE Transactions on Electron Devices*, Vol. 54, No. 1, 2007, pages 83–89. DOI: 10.1109/ted.2006.887202. 7

[62] Narang, R., Saxena, M., Gupta, R. S., and Gupta, M., Impact of Temperature Variations on the Device and Circuit Performance of Tunnel FET: A Simulation Study, *IEEE Transactions on Nanotechnology*, Vol. 12, No. 6, 2013, pages 951–957. DOI: 10.1109/tnano.2013.2276401.

[63] Datta, S., Liu, H., and Narayan, V., Tunnel FET Technology: A Reliability Perspective, *Microelectronics Reliability*, Vol. 54, 2014, pages 861–874. DOI: 10.1016/j.microrel.2014.02.002. 7

[64] Wilk, G. D., Wallace, R. M., and Anthony, J. M., High-*k* Gate Dielectrics: Current status and Materials Properties Considerations, *Journal of Applied Physics*, Vol. 89, No. 10, 2001, pages 5243–5275. DOI: 10.1063/1.1361065. 10, 58

[65] Clark, Robert D., Emerging applications For High *K* Materials in VLSI Technology, *Materials*, Vol. 7, 2014, pages 2913–2944. DOI: 10.3390/ma7042913.

[66] Guha, S. and Narayan, V., High-K/Metal Gate science and Technology, *Annual Review of Materials Research*, 2009, Vol. 39, pages 181–202. DOI: 10.1146/annurev-matsci-082908-145320.

[67] Mishra, D., Iwai, H., and Wong, H., *High-k Gate Dielectrics, Interface*, The ElectroChemical Society, 2005, pages 30–34.

[68] Andro T., Ultimate Scaling of High-*k* Gate Dielectrics: Higher *K* or Interfacial Layer Scavenging?, *Materials*, 2012, pages 478–500. DOI: 10.3390/ma5030478.

[69] Groner, M. D. and George, S. M., High-k Dielectrics Grown by Atomic Layer Deposition: Capacitor and Gate Applications, Chapter 10, *Interlayer Dielectrics for Semiconductor Technologies*, Elsevier Inc., 2003, pages 327–348. DOI: 10.1016/b978-012511221-5/50012-x. 10, 58

[70] Takagi, S., Toriumi, A., Iwase, M., and Tango, H., On The Universality of Inversion Layer Mobility in Si MOSFET's: Part I-Effects of Substrate Impurity Concentration, *IEEE Transactions on Electron Devices*, Vol. 41, No. 12, 1994, pages 2357–2362. DOI: 10.1109/16.337449. 12

[71] Labounty, C., Shakouri, A., and Bowers, J. E., Design and Characterization of Thin-Film Microcoolers, *Journal of Applied Physics*, Vol. 89, No. 7, 2001, pages 4059–4064. DOI: 10.1063/1.1353810. 15

Authors' Biographies

NABIL SHOVON ASHRAF

Nabil Shovon Ashraf was born in Dhaka, Bangladesh, in 1974. He obtained his Bachelor's degree in Electrical Engineering (EE) from Indian Institute of Technology Kanpur, India, in 1997, Master's degree in EE from University of Central Florida, Orlando, Florida, in 1999, and Doctorate degree in EE from Arizona State University, Tempe, Arizona, in 2011. He was a post doctoral researcher in the department of electrical engineering of Arizona State University from 2011–2014. He was employed as design engineer in RF Monolithics Inc., a surface acoustic wave (SAW) based filter design company in Dallas, Texas, from 1999–2001. From 2003 till 2006 he served on the faculty of the department of Electrical & Electronic Engineering of Islamic University of Technology, Gazipur, Bangladesh, as Assistant Professor. In fall 2014, he became Assistant Professor in the department of Electrical and Computer Engineering (ECE) of North South University, Dhaka, Bangladesh. He has been listed in Marquis *Who's Who in America* in 2016 (70th platinum edition) and also in 2015 (69th edition). Dr. Ashraf has, to-date, published six peer reviewed journal papers (two IEEE EDS society) and around 14 conference papers (three IEEE EDS society). He has also contributed to two book chapters on interface trap-induced reliability aspects (threshold voltage fluctuations) of nanoscale devices. He specializes in the area of device physics and modeling analysis of scaled devices for enabling improved device performance at the scaled node of current MOSFET device architectures.

SHAWON ALAM

Shawon Alam successfully completed his Bachelor's of Science in Electrical and Electronic Engineering from North South University in August 2015, and has performed research (2014–2015) concentrating on nanoscale nMOSFET device modeling on threshold voltage with the impact of temperature (100K-500K). Later, he joined an internship program at a multinational telecommunications company for a period of four months. He was involved with many extra-curricular and co-curricular activities like sports, social service, and a teaching assistantship during his graduate program. Presently, he is working as a network engineer in a renowned organization involved with international internet gateway (IIG) as well as internet service provider (ISP). In the future, he is willing to do further research on the MOSFET device.

MOHAIMINUL ALAM

Mohaiminul Alam, a distinguished graduate with a Bachelor's of Science in Electrical and Electronic Engineering from North South University, has performed research (2014–2015) concentrating on CMOS device modeling, projecting the impact of temperature (100K–500K) on threshold voltage. Later, he joined an internship program at a telecommunications company. He is currently working as a project system and development engineer in an organization involved with the research and development program of solar products and solar solutions.

Printed in the United States
by Baker & Taylor Publisher Services